Optimal Control Applications for Operations Strategy

Bowon Kim

Optimal Control Applications for Operations Strategy

Springer

Bowon Kim
KAIST Business School
Seoul
Korea (Republic of)

ISBN 978-981-10-3598-2 ISBN 978-981-10-3599-9 (eBook)
DOI 10.1007/978-981-10-3599-9

Library of Congress Control Number: 2017932003

Printed on acid-free paper

This Springer imprint is published by Springer Nature
The registered company is Springer Nature Singapore Pte Ltd.
The registered company address is: 152 Beach Road, #21-01/04 Gateway East, Singapore 189721, Singapore

For My Family

Preface

This book 'Optimal Control Applications for Operations Strategy' is about applications of optimal control theory to operations and supply chain management. While teaching masters and Ph.D. students at KAIST Business School for the last 20 years, I have found that optimal control theory is a very powerful tool to analyze and understand the fundamental issues in operations strategy. One of the most important roles played by optimal control theory is to provide managerial and economic insights, which enable the students to comprehend the dynamic activities and interactions in operations.

In the literature, however, optimal control theory is not one of the mainstream approaches to study operations management. As such, not many reference books on optimal control theory applied to operations are available. Like any other research methodology in management, it is obvious that optimal control theory alone is not complete. Nevertheless, it is certainly an effective tool to supplement other methodologies in operations, i.e., it plays a very significant role in analyzing the complex dynamics embodied in operations strategy. This book could fill the gap in the literature and contribute to complementing other research methodologies for operations and supply chain management.

It consists of five chapters, which are based on and refined versions of some of the papers I have published for the last 20 years. Each chapter starts with an abstract and keywords, followed by the 'key learning' box, which succinctly summarizes core lessons the students are expected to learn from the chapter. There are exercise problems at the end of the main text in the chapter. Detailed proofs and explanations of the theorems in the chapter appear in the appendices. I discuss the primary goals and contents for each of the chapters as follows:

Chapter 1 'Optimal Control Theory and Operations Strategy' introduces some of the basic concepts in optimal control theory and elaborates on the dynamics of production technology development, an essential part of operations strategy.

Chapter 2 'Value of Coordination in Supply Chain Management' looks into coordination as the infrastructural dimension of supply chain management, one of the most important subjects in operations strategy, and endeavors to define the value of coordination.

Chapter 3 'Innovation Competition and Strategy' discusses innovation and postulates innovation competition as a crucial factor in operations strategy, exploring the conditions under which competing firms collaborate for innovation.

Chapter 4 'Dynamic Coordination for New Product Development' puts forth that new product development calls for significant supply chain coordination and suggests how to take into account the serviceability when developing a new product.

Chapter 5 'Sustainable Supply Chain Management' identifies two key players for ensuring sustainability, i.e., the government and consumers, and examines the critical role of consumer awareness in accomplishing the environmental sustainability.

I would like to thank Mr. William Achauer in Springer Singapore for his help during the initial discussion for possible publication of my book. Bill helped me complete the proposal review process productively. My Ph.D. student Jeong Eun Sim assisted me in compiling my papers for the book. I appreciate Jeong Eun for her making a diligent effort. I also would like to acknowledge my Ph.D. students, Sunghak Kim, Taehyung Kim, Hyunjin Kang, Jaeseok Na, and Yeoyoung Cho, for their assiduous working during the proofreading.

I hope this book can help the students in operations strategy learn how to apply optimal control theory to analyze, understand, and solve actual managerial problems, especially in operations and supply chain management. I am confident that it enables the students to develop their own research capability eventually.

'I hear, I know. I see, I remember. I do, I understand.'—Confucius

KAIST, Seoul, Korea Bowon Kim
January 2017

The original version of the book was revised. For detailed information please see erratum. The erratum to the book is available at 10.1007/978-981-10-3599-9_6

Contents

Chapter 1
Optimal Control Theory and Operations Strategy

Abstract In this chapter, I introduce basic concepts of optimal control theory. Then, I discuss dynamics of production technology development and show how to apply optimal control theory to answering an important operations strategy issue, i.e., how to develop production technology dynamically.

Keywords Optimal control theory · Production technology · Manufacturing learning · Operations strategy

Key Learning

- *What is optimal control theory?*
  - Optimal control theory is an analytical methodology to solve dynamic optimization problems in economics and management.
  - We define the optimal control theory model and elaborate the Pontryagin's maximum principle to solve an optimal control theory problem.
- *How should the firm allocate resources to develop production technology?*
  - We consider two types of mechanism to develop production technology, i.e., endogenous (in-house or on-shop) and exogenous (subcontracting or off-shop).
  - Manufacturing learning propensity espoused by managers toward a particular technological direction is a significant factor to shape the attention allocation dynamics.
  - Learning propensity developed by the management for a certain type of production technology is determining the optimal dynamics of technology development.

The original version of this chapter was revised. The erratum to this chapter is available at 10.1007/978-981-10-3599-9_6

B. Kim, *Optimal Control Applications for Operations Strategy*,
DOI 10.1007/978-981-10-3599-9_1

- Small decisions accumulated by complying with a benign choice of particular learning propensity at the early stage might become an irreversible force the firm could not deny to follow in the later stage.
- Top management must be able to understand the early formation of learning propensity by managers, in order to optimally manage the dynamics of technology development.

Optimal control theory is an analytical methodology to solve dynamic optimization problems in economics and management. Optimization is a goal in economic and managerial decision-making. For instance, a firm might have such goals as revenue or market share maximization, profit maximization, cost minimization, project lead-time minimization, yield maximization, customer satisfaction maximization, and inventory minimization. An economic or managerial decision-making involves many factors, i.e., variables and parameters. Parameters are in general given or defined by environmental forces. The decision-maker deals with two different types of variables, decision (control) variables, and state variables. A decision or control variable represents a specific decision, the decision-maker should make in order to accomplish her goal, e.g., how many units to produce, how much to invest in marketing or R&D, how many workers to hire, and how much to charge the customer for the product. A state variable represents the current state of a certain measure, resulting from the decision made by the decision-maker. For instance, the current level of the firm's market share is a state variable, resulting from the decisions made by the firm in the previous periods, e.g., decisions on production amount, marketing expenditure, price, and R&D.

1 Basics of Optimal Control Theory

In this section, we define the optimal control theory model and elaborate the most important methodology, i.e., Pontryagin's maximum principle (Kamien and Schwartz 1991), to solve an optimal control theory problem. Our objective here is not to put forth a detailed theoretical derivation of the maximum principle, but to teach the student a working knowledge about it. If the student is interested in learning more theoretical aspects of optimal control theory, she should also study some of the theory books on optimal control theory.

1.1 Optimal Control Theory Model

An optimal control theory model is obtained by introducing the control variable into a continuous-time dynamic optimization model. Consider an objective function as follows: $Max \int_0^T G(t, y(t), u(t))dt$ where $u(t) = y'(t)$ is a control variable. Noting that $y(t)$ is a state variable, we need to define the dynamic behavior of the state variable in terms of the control variable, i.e., to express $y'(t)$ as a function of $u(t)$ and $y(t)$: $y'(t) = f(t, y(t), u(t))$. Assuming that the relevant time horizon is $t_0 \le t \le t_1$, we present a standard form of an optimal control theory problem:

$$Max \int_{t_0}^{t_1} G(t, y(t), u(t))dt$$

$$\begin{aligned} Subject\ to\quad & y'(t) = f(t, y(t), u(t)) \\ & y(t_0) = y_0, t_0, t_1 \text{ fixed (given), and } y(t_1) \text{ free (unfixed).} \end{aligned}$$

1.2 Maximum Principle

Lev Semyonovich Pontryagin developed the maximum principle, which postulates necessary conditions to solve optimal control theory problems. We summarize the necessary conditions based on Pontryagin's maximum principle for the optimal control theory model as follows:

$$Max \int_{t_0}^{t_1} f(t, x(t), u(t))dt \tag{1}$$

$$\begin{aligned} Subject\ to\quad & x'(t) = g(t, x(t), u(t)) \\ & x(t_0) = x_0, t_0, t_1 \text{ fixed (given), and } x(t_1) \text{ free (unfixed)} \end{aligned}$$

First, we form the Hamiltonian:

$$H(t, x(t), u(t), \lambda(t)) = f(t, x(t), u(t)) + \lambda(t)g(t, x(t), u(t)) \tag{2}$$

Then, the necessary conditions are as follows:

$$\frac{\partial H}{\partial u} = f_u + \lambda g_u = 0, \tag{3}$$

$$\lambda' = -\frac{\partial H}{\partial x} = -f_x - \lambda g_x, \tag{4}$$

$$x' = \frac{\partial H}{\partial \lambda} = g, \tag{5}$$

$$\lambda(t_1) = 0 \text{ if } x(t_1) \text{ free (transversality condition),} \tag{6}$$

$$x(t_0) = x_0 \text{ (initial condition)}. \tag{7}$$

In addition, we need

$$\begin{aligned} H_{uu}(t, x^*, u^*, \lambda) \le 0 \text{ for maximization and} \\ H_{uu}(t, x^*, u^*, \lambda) \ge 0 \text{ for minimization.} \end{aligned} \tag{8}$$

Exercise 1 The revenue of a firm increases with its capability, $x(t)$, which in turn can be enhanced by $u(t)$, the level of reinvestment of firm's capability. The company formulates its decision problem as follows:

$$\textit{Maximize} \int_0^4 (5x - u^2)\mathrm{d}t \tag{9}$$

$$\begin{aligned} \textit{Subject to} \quad & \dot{x} = u \\ & x(0) = 0, \\ & x(4) \text{ free.} \end{aligned}$$

Find the optimal reinvestment policy and the stock of firm's capability that maximizes the total profit of the firm over $0 \le t \le 4$.

Solution:

First, form the Hamiltonian:

$$H = 5x - u^2 + \lambda u \tag{10}$$

Using (3), we take $H_u = -2u + \lambda = 0$, $\therefore u = \frac{1}{2}\lambda$.

According to (4), $\lambda' = -H_x = -5$. Now $\lambda = -5t + c$, where c is a constant to be determined. Since $\lambda(4) = 0$ due to the transversality condition, $\lambda(4) = -5 \times 4 + c = 0$ and $c = 20$. $\therefore \lambda = -5t + 20$.

$$\therefore u = \frac{1}{2}\lambda = -\frac{5}{2}t + 10.$$

Since $\dot{x} = u$, $\dot{x} = -\frac{5}{2}t + 10$. Therefore, $x(t) = -\frac{5}{4}t^2 + 10t + k$ where k is a constant to be determined. Since $x(0) = 0$ due to the initial condition, $x(0) = k = 0$. $\therefore x = -\frac{5}{4}t^2 + 10t$.

Since $5x - u^2 = 5\left(-\frac{5}{4}t^2 + 10t\right) - \left(-\frac{5}{2}t + 10\right)^2 = -\frac{25}{2}t^2 + 100t - 100$, we have $\int_0^4 (5x - u^2)\mathrm{d}t = \int_0^4 \left(-\frac{25}{2}t^2 + 100t - 100\right)\mathrm{d}t = \frac{400}{3}$.

We can draw the optimal paths of the variables (Fig. 1) and also the objective function (Fig. 2).

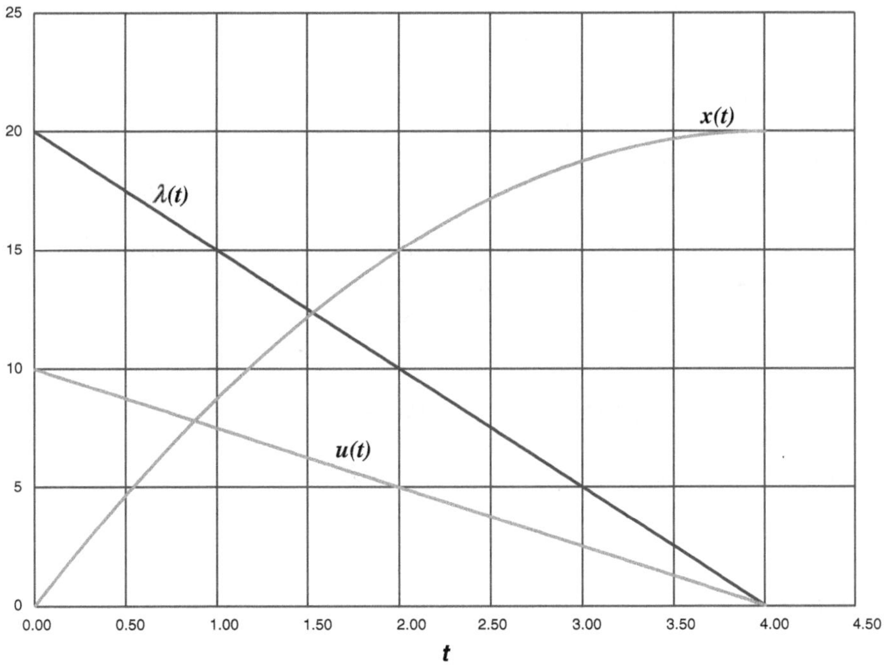

Fig. 1 Dynamic paths of variables

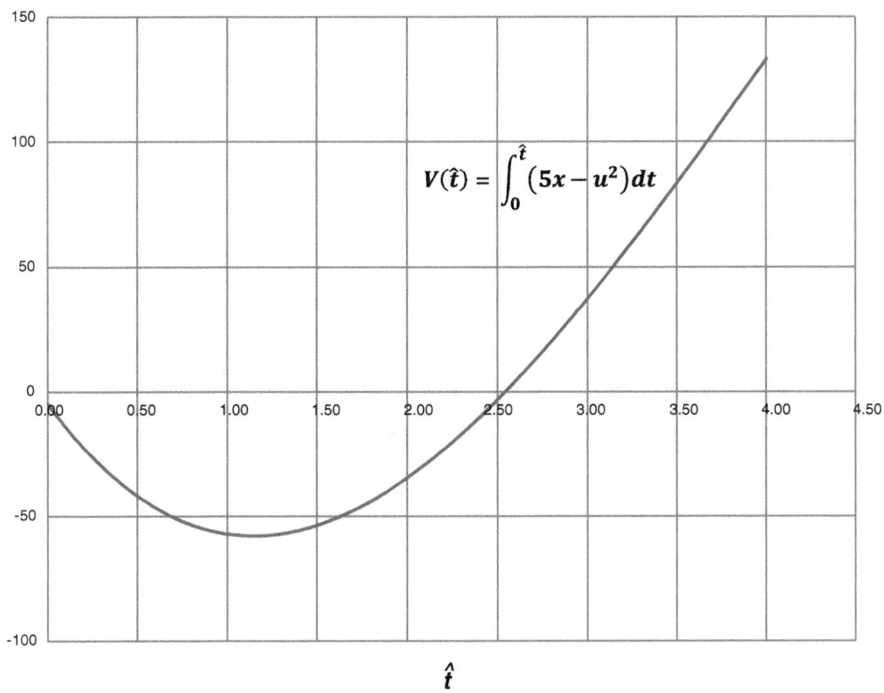

Fig. 2 Dynamic path of objective function

2 Dynamics of Production Technology Development[1]

In this section, we explore the impact of manufacturing learning on the dynamics of production technology development. Let us consider two types of mechanism to develop production technology: endogenous (in-house or on-shop) and exogenous (subcontracting or off-shop) in relation to the '*manufacturing process.*' Endogenously developed technology is more firm-specific, conducive to in-line learning within the manufacturing process, less uncertain in development success than exogenously developed technology. How to allocate managerial resources (also, attention or commitment) between on-shop and off-shop development efforts can be determined by the learning intention or propensity possessed by the managers.

In the relatively early stage of decision time horizon, the managers participating in the development process form learning propensity for either on-shop technology or off-shop technology, make decisions, and behave accordingly to the formed propensity: Learning is assumed to affect the implementation cost. Since in the early period the benefit from technological improvement outweighs the implementation cost, the learning impact seems minimal. As the terminal point of decision horizon approaches, the implementation cost becomes dominant over the benefit (since the benefit decreases more rapidly than the cost does). The firm may eventually face a time point at which the '*learning-induced*' bias to utilize the particular '*chosen mechanism*' to develop technology becomes so enormous that the firm cannot control its own dynamics with minor adjustment: The dynamics of the later period was already prophesied by the initial choice of learning propensity (*freezing effect of learning intention*; *catastrophic effect of learning prophecy*).

Top management should understand the early formation of *learning propensity by (middle) managers* in order to optimally control the dynamics of production technology development.

2.1 Introduction

Production knowledge is one critical factor for manufacturing improvement. Arrow (1969) posited that a typical firm is engaged in activities ranging from the production of pure knowledge to the production of pure product. Economists have regarded '*R&D (research and development)*' function as a primary source of technological progress in a rather broad context. It is, however, too vague and rigid to be applicable for the research on production at the firm level. Embracing the manufacturing process as a source of production knowledge development, Jaikumar

[1]This section is a refined version of Kim, B. (1996) 'Learning-induced control model to allocate managerial resources for production technology development.' *International Journal of Production Economics, 43* (2–3), 267–282.

and Bohn (1992) offered three ways to generate production knowledge: (i) purchasing outside knowledge (external to the firm), (ii) conducting intensive R&D outside manufacturing (internal to the firm), and (iii) learning within existing manufacturing (endogenous to the firm).

Arrow (1962) opened up the avenue of production learning theory by stylizing the concept of 'learning by doing.' Many studies have focused on the learning theory (Alchian 1963; Yelle 1979; Sáenz-Royo and Salas-Fumás 2013). Despite the abundance of economic studies on learning phenomena, we find it inadequate for most of them to consider 'learning' as a by-product unintentionally grown out of a capital investment process. Researchers have tried to understand '*learning phenomenon*' from within the manufacturing process perspective (Fine 1986; Tapiero 1987; Bohn 1988; Dorroh et al. 1994; Xiao and Gaimon 2013). We espouse the philosophy that the process to generate production knowledge consists of conscious efforts exerted by the firm engaging in intelligent decision-making on the balance between endogenous and exogenous mechanisms.

2.2 Production Technology Development

Manufacturing technology is technology of process control (Jaikumar 1988). Dosi (1982) defined technology as a set of pieces of knowledge, both directly 'practical' (related to concrete problems and devices) and 'theoretical' (but practically applicable although not necessarily already applied), know-how, methods, procedures, experience of successes and failures, and physical devices and equipment: In effect, the production technology is 'procedural knowledge' about manufacturing process. We can abstract two defining attributes of production technology: (i) It is a set of knowledge, procedures, and devices, both conceptual and physical and (ii) it can be improved by utilizing the mechanism of problem-solving activity in manufacturing process. We suggest two types of mechanism to develop production technology: on-shop mechanism and off-shop mechanism, which can be compared to '*local search process*' and '*imitation process*,' respectively (Nelson and Winter 1973).

Endogenous (on-shop) mechanism is strictly confined within the actual production system itself, not dominated by nonmanufacturing supplementary functions in the firm. For instance, new production knowledge solely developed in the R&D department in a firm may not be regarded as endogenous since it is neither initiated nor guided by the factory environment. There is, however, a possibility of production knowledge development through the collaboration between factory engineers and R&D scientists: As long as a substantive role in the production knowledge development is played at the factory level, we can regard that effort as endogenous. The concept of endogenous production knowledge development can be analogized with the coevolution resulting from 'running the factory as learning laboratory (Leonard-Barton 1992).'

Exogenous (off-shop) mechanism is the contrasting concept to the endogenous one. Most of the environmental constituents surrounding the manufacturing factory can be potential sources of exogenous knowledge development. Research institutes, probably external to the manufacturing system, may offer new technological breakthroughs to the firm. As alluded above, production knowledge solely developed by nonmanufacturing functions such as in-house R&D activity shall be considered as exogenously sourced.

We can assess the potential of technology mechanism according to various attributes of production technology (Table 1): The potential is concerned with each mechanism's effectiveness as means to develop or improve production technology. Note that Table 1 is just a general example of each mechanism's potential, whose exact level might depend on specifics faced by each manufacturing system in the real-world context. In general, for example, '*high firm-specificity*' is associated with an '*extensive*' potential for '*on-shop (endogenous) mechanism*.' The same logic can be applied to excludability. An opposite argument can be made for transferability and fidelity, although the fidelity issue less concerns the on-shop production technology: If there exists high fidelity, the potential of an off-shop mechanism can be extensive (i.e., the off-shop mechanism can be very effective), since it is not dampened by the lack of fidelity. In-line learning relates to endogenous technology,

Table 1 Potential of each technology mechanism

Production technology and knowledge attributes	Endogenous mechanism (on-shop)		Exogenous mechanism (off-shop)	
Composite measure	Limited	Extensive	Limited	Extensive
Rivalry	L	H	H	L
Excludability	L	H	H	L
Firm-specificity (system embeddedness: human and physical)	L	H	H	L
Transferability	H	L	L	H
Fidelity	H*	L*	L	H
In-line learning	L	H	H*	L*
Remote learning	H*	L*	L	H
Industry-wide standardization (regarding technology)	H	L	L	H
Industry-wide standardization (regarding demand preference)	L	H	H	L
Technology obsolescence (knowledge decaying)	Fast	Slow	Slow	Fast
Implementation and adaptation velocity[a]	L	HH	L	H
Development velocity[b]	L	H	L	HH
Development uncertainty[c]	H	LL	H	L
Development continuity[d]	L	HH	L	H

Note (i) H means 'high' and L means 'low.' (ii) '*' implies 'only insignificantly related'
[a]Higher in endogenous than in exogenous
[b]Higher in exogenous than in endogenous
[c]Lower in endogenous than in exogenous
[d]Higher in endogenous than in exogenous

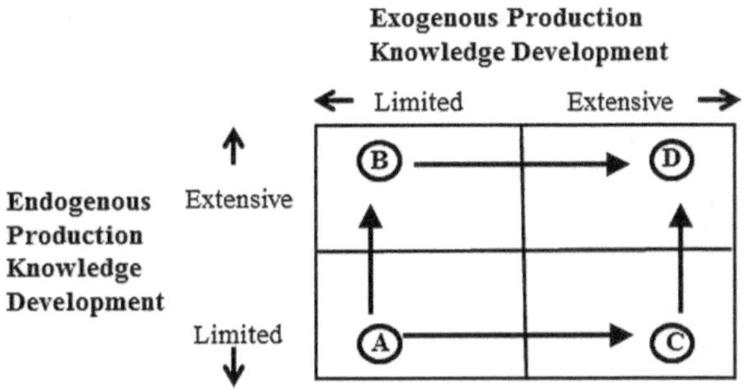

Fig. 3 Framework of analyzing production technology development

whereas remote learning relates to exogenous technology. If the possibility of in-line learning is very high, then the potential of endogenous knowledge development can be extensive. Similar points can be made for other attributes. The last four attributes are more related to comparison between 'endogenous' and 'exogenous' than between 'limited' and 'extensive': For instance, endogenous knowledge development requires much less adaptation (to the existing manufacturing system) period than exogenous knowledge development probably does.

The framework based on technology mechanisms (Fig. 3) illustrates a typology to analyze production knowledge development. For an illustration purpose, Fig. 3 contains plausible dynamic paths a firm might go through over time. One possible evolution might start from position 'A' to 'B': from limited endogenous and limited exogenous knowledge development potential to extensive endogenous development with limited exogenous development potential. This move can be made if the firm has invested in enhancing its infrastructure conducive to endogenous development, or if the operating context given to the firm has changed toward more favorable environment for endogenous development. Through either an intentional effort or a situational change, the firm might be able to move to a more productive position, 'D' where both endogenous and exogenous development potentials are extensive. Another example shows a path from 'A' to 'C', then eventually 'D'. Some other possible routes can be constructed: We will show that the logic behind the move has a metaphor in economics, '*induced bias hypothesis*' (Hicks 1932; Feller 1972).

2.3 Dynamic Optimal Control Model

This section establishes a dynamic optimal control model: The mathematical analysis draws conceptual conditions under which particular dynamics to improve production technology can be more effective than the others. A relevant dynamic optimal control model must distinguish endogenous production technology from exogenous technology. The modeling will take into account those attributes, scrutinized in the previous sections, of production knowledge.

$N(t)$: resources spent for endogenous technology development at time t,
$X(t)$: resources spent for exogenous technology development at time t,
$N(t)$, $X(t)$: on-shop and off-shop efforts, respectively, at basic capability level,
$A_N(t)$: endogenously developed knowledge or technology (control capability) level at t,
$A_X(t)$: exogenously developed knowledge or technology (control capability) level at t,
$S(t)$: state variable (system capability) such as production flexibility or lead time at t,
$U(S, N, X, A_N, A_X, t)$: utility function for the production system at t.

The model assumes that a system level capability such as production lead time or productivity can be improved through either on-shop or off-shop technology (control capability): Managerial resources (attention) allocated for on-shop ($N(t)$) and off-shop ($X(t)$) activities improve on-shop technology ($A_N(t)$) and off-shop technology ($A_X(t)$), respectively, which in turn improve the system level productivity ($S(t)$).

2.3.1 Dynamic Equations of Production Technology as Control Capability

Among the production knowledge attributes, there are transferability, fidelity, and in-line learning. All of these relate to the degree of successful development of production technology, the uncertainty attribute embedded in the production knowledge development: A dynamic equation of production knowledge needs to have a coefficient to express these attributes. It also requires a coefficient associated with the process to convert the resource unit into the knowledge unit, a coefficient of 'returns to scale.' Assuming that the unit conversion is a relationship of nonlinear function, we use a unit conversion coefficient in an exponential form. Shell (1966) modeled the growth in the stock of technical knowledge with a differential equation containing a term of 'technical knowledge decay,' connected with such knowledge attributes as firm-specificity and technological obsolescence: Unless the firm reinforces and reutilizes the developed production knowledge, the overall stock of knowledge will decay over time (Gaimon et al. 2011):

$$\dot{A}_N(t) = \sigma_N(t)[N(t)]^{\alpha_N(t)} - \rho_N(t)A_N(t), \tag{11}$$

$$\dot{A}_X(t) = \sigma_X(t)[X(t)]^{\alpha_X(t)} - \rho_X(t)A_X(t), \tag{12}$$

where $\sigma_i(t)$ is the coefficient (probability) of development success at t, $0 \le \sigma_i(t) \le 1, i = N\,or\,X, \alpha_i(t)$ the coefficient of unit conversion (returns to scale), $0 \le \alpha_i(t) \le 1$, and $\rho_i(t)$ the production knowledge decaying rate, $0 \le \rho_i(t) \le 1$. We further assume that

$$\sigma_N(t) > \sigma_X(t), \alpha_X(t) > \alpha_N(t) \; and \; \rho_N(t) < \rho_X(t). \tag{13}$$

In other words, (i) other things being equal, it is easier (more successful) to develop endogenous production knowledge than to import exogenous production knowledge for each unit resource spent, (ii) off-shop return to scale is larger than on-shop return to scale, and (iii) on-shop production technology is more firm-specific or stickier than off-shop production technology (Von Hippel 1994), i.e., endogenous knowledge decays more slowly than exogenous knowledge: This observation relates to the knowledge 'tacitness' as in Mody (1989).

Let *R(t)* represent the available managerial resources at *t*: The firm can decide how much managerial attention or resources to devote to either on-shop or off-shop technology development effort. The resource constraint is as follows:

$$N(\mathrm{t}) + X(t) \leq R(t). \tag{14}$$

2.3.2 Dynamic Equation of System Dimension as System Capability

We assume simple 'exponential nonlinearity' between production knowledge (control capability) and system capability: A linear format is just a special case of the generic nonlinear formulation. We further assume the additivity of production knowledge types, and the system capability's decaying as time passes on:

$$\dot{S}(t) = \beta_N(t)[A_N(t)]^{\gamma_N(t)} + \beta_X(t)[A_X(t)]^{\gamma_X(t)} - \phi(t)S(t), \tag{15}$$

where $\beta_i(t)$ is the coefficient of production knowledge's contribution to system capability, $\gamma_i(t)$ the 'returns to scale' of production technology, and $\phi(t)$ the system capability decaying rate, $0 \leq \phi(t) \leq 1$. We also assume that

$$\beta_N(t) < \beta_X(t) \quad \text{and} \quad 1 \leq \gamma_N(t) \leq \gamma_X(t) \tag{16}$$

2.3.3 System Utility Function

The objective function of the optimal control problem is determined by the market choices. A reasonably acceptable type is a linear equation (Dompere 1993), assuming the linear equation is valid in the relevant time span,[2] [0, T]:

[2]There are several reasons why the firm may have a finite time period for a specific system capability: (i) first of all, there is a possibility that the market preference with regard to a product dimension might change over time; for instance, at one point the customers might think of 'low production cost' as a major buying determinant, but after a certain period of time, they might emphasize 'short production lead time' over 'production cost' as a critical decision factor, and (ii) the firm's infrastructure might be conditioned by the structural factors; it is conceivable that if the fundamental structural conditions change, then the entire meaning of infrastructural mechanism

$$U(S,N,X,A_N,A_X,t) = \psi(t)S(t) - C_N(t)N(t) - C_X(t)X(t), \tag{17}$$

where $\psi(t)$ is the market value of system capability, $S(t)$ at t, $C_i(t)$ the market (opportunity) cost to utilize a unit of resource type $i, i = N$ or X, and

$$C_N(t) \leq C_X(t), \tag{18}$$

which implies that the unit implementation cost of an on-shop mechanism is less expensive than that of an off-shop mechanism. The cost is associated with an opportunity cost of the resources plus an additional implementation cost necessary to apply each resource unit for developing each type of production technology.

2.3.4 Model Solution

For simplicity and traceability, we make further assumptions without loss of generality. First, we assume that the parameters are time-invariant within [0, T]: Thus, $\psi(t) = \psi$, $C_i(t) = C_i$, $\beta_i(t) = \beta_i$, so on and also $R(t) = R$, a given amount, for $t \in [0, T]$. Second, supposing that $A_N(t)$ and $A_X(t)$ are already converted to the same unit as that of $S(t)$ through $\alpha_N(t)$ and $\alpha_X(t)$, we assume $\gamma_i(t) = 1$ in the ensuing model. Finally, we normalize the given human resources so that we can use $N(t)$ and $X(t)$ as ratios rather than absolute amounts. Assuming that R is utilized only for the production knowledge development, we express $1 = X(t) + N(t)$. Now, we have a dynamic control problem as follows, i.e., DCP1:

[DCP1]

$$\mathit{Maximize} \int_0^T (\psi S(t) - C_N RN(t) - C_X(R - RN(t)))e^{-rt}\mathrm{d}t$$

$$\mathit{Subject\ to}\ \dot{S}(t) = \beta_N A_N(t) + \beta_X A_X(t) - \phi S(t) \tag{19}$$

$$\dot{A}_N(t) = \sigma_N [RN(t)]^{\alpha_N} - \rho_N A_N(t) \tag{20}$$

$$\dot{A}_X(t) = \sigma_X [R - RN(t)]^{\alpha_X} - \rho_X A_X(t) \tag{21}$$

$$0 \leq N(t) \leq 1 \tag{22}$$

$$A_N(0) = A_{N0}, \quad A_X(0) = A_{X0}, \quad \mathrm{S}(0) = S_0$$

$$A_N(T),\ A_X(T),\ \text{and}\ \mathrm{S}(T)\ \text{are free.} \tag{23}$$

(Footnote 2 continued)

is put in a different context; therefore, [0, T] corresponds to the period during which the structural condition remains the same.

The current value Hamiltonian[3] of (DCP1) is

$$\begin{aligned} H_C = & (\psi S(t) - (C_N - C_X)RN(t) - C_X R) \\ & + m_1(\beta_N A_N(t) + \beta_X A_X(t) - \phi S(t)) \\ & + m_2(\sigma_N R^{\alpha_N}[N(t)]^{\alpha_N} - \rho_N A_N(t)) \\ & + m_3(\sigma_X R^{\alpha_X}[1 - N(t)]^{\alpha_X} - \rho_X A_X(t)), \end{aligned}$$

where $m_i = \lambda_i e^{rt}$ with λ_i the costate variable for associated differential equation.[4]

The current value Lagrangian[5] is $\mathcal{L}_C = H_C + n(1 - N(t))$ where $n = \theta e^{rt}$, with θ the Lagrangian multiplier for (22).

The optimality condition requires

$$\frac{\delta \mathcal{L}_C}{\delta N} = -(C_N - C_X)R + m_2 \sigma_N \alpha_N R^{\alpha_N}[N(t)]^{\alpha_N - 1} - m_3 \sigma_X \alpha_X R^{\alpha_X}[1 - N(t)]^{\alpha_X - 1} - n = 0, \tag{24}$$

$$\frac{\delta \mathcal{L}_C}{\delta n} \geq 0, n^*(t) \geq 0, n^*(t)[1 - N^*(t)] = 0. \tag{25}$$

The sufficient condition calls for

$$\begin{aligned} \frac{\delta^2 \mathcal{L}_C}{\delta N^2} = & m_2 \sigma_N \alpha_N (\alpha_N - 1) R^{\alpha_N}[N(t)]^{\alpha_N - 2} \\ & + m_3 \sigma_X \alpha_X (\alpha_X - 1) R^{\alpha_X}[1 - N(t)]^{\alpha_X - 2} \leq 0. \end{aligned} \tag{26}$$

Applying the *maximum principle* by Pontryagin (1962), we can obtain the optimal paths for costate variables: Detailed derivation procedure is in the appendix.[6]

$$\lambda_1^* = \frac{\Psi e^{-rt}}{r + \phi}\left\{1 - e^{-(\phi + r)(T - t)}\right\}, \tag{27}$$

$$\begin{aligned} \lambda_2^* = & \frac{\beta_N \Psi e^{-rt}}{(\phi - \rho_N)(r + \rho_N)(r + \phi)}\Big\{(\mathrm{r} + \rho_N)e^{-(r + \phi)(T - t)} \\ & - (r + \phi)e^{-(r + \rho_N)(T - t)} + (\phi - \rho_N)\Big\}, \end{aligned} \tag{28}$$

---

[3]Refer to Appendix 3.

[4]r is an internal discount rate.

[5]Refer to Appendix 4.

[6]Since R is not constrained in the formulation, it can be regarded as an externally determined constant. The model implicitly presumes that the parameters are within the 'relevant' range so that optimal dynamics can exist.

$$\lambda_3^* = \frac{\beta_X \Psi e^{-rt}}{(\phi - \rho_X)(r + \rho_X)(r + \phi)} \Big\{ (\mathrm{r} + \rho_X) e^{-(r+\phi)(T-t)} \\ -(r+\phi) e^{-(r+\rho_X)(T-t)} + (\phi - \rho_X) \Big\}. \tag{29}$$

Observation 1 *$N(t)^*$ which solves* (24) *and* (25) *is an optimal solution which maximizes* [DCP1] *with given constraints (sufficient condition for optimality).*

Proof We need to show that m_2 and m_3 (λ_2^* and λ_3^*) are always positive. First, we will prove that λ_2^* is always positive. From (28), $\lambda_2^*(T) = 0$. If it is true that $\dot{\lambda}_2^* \leq 0$, then $\lambda_2^* \geq 0$ always.

From (28), $\dot{\lambda}_2^* = \frac{\beta_N \psi}{(r+\rho_N)(r+\phi)} \left\{ \phi \frac{r+\rho_N}{\phi - \rho_N} e^{-(r+\phi)T} \mathrm{e}^{\phi t} - \rho_N \frac{r+\phi}{\phi-\rho_N} e^{-(r+\rho_N)T} e^{\rho_{N_t}} - re^{-rt} \right\}$.

Thus, $\dot{\lambda}_2^* \leq 0$ is equivalent to $\phi \frac{r+\rho_N}{\phi - \rho_N} e^{-(r+\phi)T} e^{\phi t} - \rho_N \frac{r+\phi}{\phi-\rho_N} e^{-(r+\rho_N)T} e^{\rho_{N_t}} - re^{-rt} \leq 0$ or

$$\phi \frac{r+\rho_N}{\phi - \rho_N} e^{-(r+\phi)(T-t)} - \rho_N \frac{r+\phi}{\phi - \rho_N} e^{-(r+\rho_N)(T-t)} - r \leq 0, \tag{30}$$

$$-r = \rho_N \frac{r+\phi}{\phi - \rho_N} - \phi \frac{r+\rho_N}{\phi - \rho_N}.$$

Let $\mathrm{LH}(\rho_N) \equiv \frac{\phi(r+\rho_N)}{\rho_N(r+\phi)}$ and $\mathrm{RH}(\rho_N) \equiv \frac{1-e^{-(r+\rho_N)(T-t)}}{1-e^{-(r+\phi)(T-t)}}$.

Then, (30) $\Leftrightarrow \mathrm{LH}(\rho_N) \geq \mathrm{RH}(\rho_N)$ if $\phi \geq \rho_N$, or $\mathrm{LH}(\rho_N) < RH(\rho_N)$ if $\phi < \rho_N$: If $\phi = \rho_N$, then $\mathrm{LH}(\rho_N) = \mathrm{RH}(\rho_N)$. Let us consider the three values of ρ_N as given in Table 2 (see also Fig. 4).

It is trivial to show $\phi \frac{(r+1)}{(r+\phi)} \leq 1$ and $\frac{1-e^{-r(T-t)}}{1-e^{-(r+\phi)(T-t)}} \leq 1 \leq \frac{1-e^{-(r+1)(T-t)}}{1-e^{-(r+\phi)(T-t)}}$, and $\frac{\delta \mathrm{LH}}{\delta \rho_N} \leq 0$ and $\frac{\delta^2 \mathrm{LH}}{\delta \rho_N^2} \geq 0$, $\frac{\delta \mathrm{RH}}{\delta \rho_N} \geq 0$ and $\frac{\delta^2 \mathrm{RH}}{\delta \rho_N^2} \leq 0$.

It is proved that (30) $\Leftrightarrow \mathrm{LH}(\rho_N) \geq \mathrm{RH}(\rho_N)$ if $\phi \geq \rho_N$, or $\mathrm{LH}(\rho_N) < RH(\rho_N)$ if $\phi < \rho_N$. Therefore, $\dot{\lambda}_2^* \leq 0$ and $\lambda_2^*(t) \geq 0$.

Similarly, we can show $\lambda_1^* \geq 0$ and $\lambda_3^* \geq 0$ always.

Suppose the decision-maker (say, the firm) can adjust the standard units related to the decision variables (i.e., system capability variable, control capability variable, and resource variables) along the associated parameters (i.e., coefficients) in the

Table 2 Changes in $\mathrm{LH}(\rho_N)$ and $\mathrm{RH}(\rho_N)$ as ρ_N changes

ρ_N	0	ϕ	1	With regard to ρ_N
$\mathrm{LH}(\rho_N)$	∞	1	$\frac{\phi(r+1)}{(r+\phi)}$	Decreasing convex
$\mathrm{RH}(\rho_N)$	$\frac{1-e^{-r(T-t)}}{1-e^{-(r+\phi)(T-t)}}$	1	$\frac{1-e^{-(r+1)(T-t)}}{1-e^{-(r+\phi)(T-t)}}$	Increasing concave

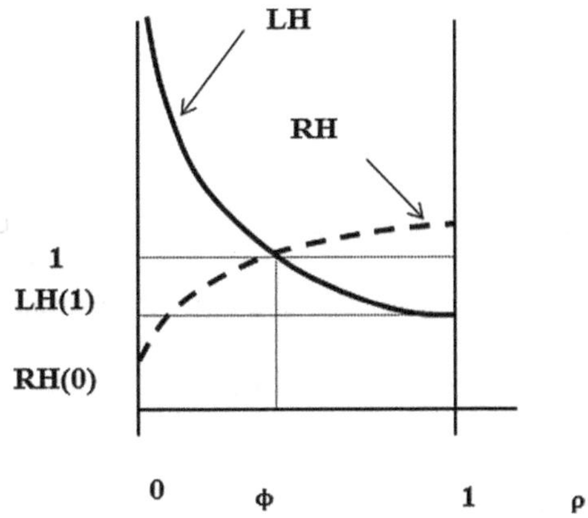

Fig. 4 Graphic changes in $\mathrm{LH}(\rho_N)$ and $\mathrm{RH}(\rho_N)$ as ρ_N changes

model so that a unit resource spent for an off-shop mechanism can be converted to one unit of off-shop production technology (if the success rate of the conversion is 1), then we can normalize $\alpha_X = 1$ without loss of generality.

Given the normalized value of $\alpha_X = 1$, the optimal path for $N^*(t)$ is

$$N^*(t) = \left[\frac{m_2\sigma_N\alpha_N R^{\alpha_N}}{(C_N - C_X)R + m_3\sigma_X R + n}\right]^{1/(1-\alpha_N)}. \tag{31}$$

2.3.5 Value Interpretation of the Nonlearning Control Model

$N^*(t)$ is the proportion of resources (i.e., managerial attention available for production technology development) to be allocated for the on-shop floor activities (endogenous mechanism). How can we interpret $N^*(t)$ in a more managerial way?

Nonlimiting Process ($\mathrm{T} < \infty$)

From (31), we know that dealing with $N^*(t)$ up to the point $N^*(t) = 1$ (i.e., while $n = 0$),

$$N^*(t) = \frac{1}{R}\left[\frac{m_2\sigma_N\alpha_N}{C + m_3\sigma_X}\right]^{1/(1-\alpha_N)} = \frac{1}{R}\left[\frac{\lambda_2^*\sigma_N\alpha_N}{Ce^{-rt} + \lambda_3^*\sigma_X}\right]^{1/(1-\alpha_N)}. \tag{32}$$

Costate variables (like λ_1^*, λ_2^*, and λ_3^*) represent the shadow prices of their respective production technologies: For instance, λ_2^* is the increase of optimal objective value if there is a unit increase in 'on-shop technology' at t. The following interpretations can be made for the terms in either numerator or denominator of $N^*(t)$.

(1) Price interpretation

Ce^{-rt}: Present value of $(C_N - C_X)$, which is *off-shop mechanism*'s (implementation) cost advantage over *on-shop mechanism*'s.

$\lambda_3^* \sigma_X$: '*Off-shop*' net contribution to the objective function; only the successful proportion (reflected in σ_X, which is the development success rate of off-shop mechanism) of shadow value can be regarded as net contribution.

$Ce^{-rt} + \lambda_3^* \sigma_X$: Total value provided by '*off-shop*' technology (considering the implementation cost differential with on-shop technology).

$\lambda_2^* \sigma_N \alpha_N$: 'On-shop' net contribution to the objective function; total value provided by 'on-shop' technology (α_N plays as a scale factor).

$\lambda_2^* \sigma_N \alpha_N / (Ce^{-rt} + \lambda_3^* \sigma_X)$: Value ratio of 'on-shop' technology to 'off-shop' technology.

Thus, the proportion of resources allocated for on-shop mechanism is the ratio between the values provided by on-shop and off-shop technologies: the larger the value provided by the on-shop technology (in comparison with the off-shop technology), the larger the $N^*(t)$. Since the shadow values change over time, we can expect $N^*(t)$ to evolve dynamically.

As long as $C_N - C_X < 0$, an extreme case might arise where there would be a point (before T) at which $Ce^{-rt} + \lambda_3^* \sigma_X < 0$ (since C is time-invariant whereas λ_2^* and λ_3^* are decreasing as time passes), and thus $N^* = 1$: A more general situation is that $Ce^{-rt} + \lambda_3^* \sigma_X < \lambda_2^* \sigma_N \alpha_N$ before $Ce^{-rt} + \lambda_3^* \sigma_X < 0$. Shortly before N^* becomes 1, there would be a period of accelerating rate of increase in N^*. This is an intuitive outcome: (i) $Ce^{-rt} + \lambda_3^* \sigma_X < \lambda_2^* \sigma_N \alpha_N$ implies that the total value of on-shop technology is larger than that of off-shop technology, (ii) $Ce^{-rt} + \lambda_3^* \sigma_X < 0$ means the shadow (total) value of off-shop technology is negative; therefore, there is no reason to spend resources for the exogenous technology development (say, $1 - N^*(t) = 0$). $Ce^{-rt} + \lambda_3^* \sigma_X < 0$ can be rearranged so that $C_N e^{-rt} + \lambda_3^* \sigma_X < C_X e^{-rt}$, implying that the present value of cost associated with the off-shop technology is larger than the total contribution provided by the off-shop technology (is equal to the present value of on-shop implementation cost plus the net shadow value of off-shop technology): When the cost is larger than the benefit, the firm should not allocate any resources for the off-shop technology development.

As long as $C_N - C_X > 0$, there would be a point (before T) at which $\lambda_2^* \sigma_N \alpha_N / (Ce^{-rt} + \lambda_3^* \sigma_X) \cong 0$ (since $Ce^{-rt} > 0$ while it must be that $\lambda_2^*(T) = \lambda_3^*(T) = 0$), and thus, $N^* = 0$. Shortly before N^* becomes 0, there would be a period of accelerating rate of decrease in N^*. If the (shadow) value of on-shop technology remains sufficiently large throughout the decision time horizon, it might never occur that $N^* = 0$. Nevertheless, we can expect N^* to decrease as the terminal time approaches.

(2) Marginal value interpretation.

Here, we proceed in a direction opposite to the one adopted for the previous derivation: We start with a plain question, '*What is the value afforded by each production technology we can guess before ever calculating N^*?*'

From (31), supposing $n = 0$,

$$N^*(t) = \left[\frac{m_2\sigma_N\alpha_N R^{\alpha_N}}{(C_N - C_X)R + m_3\sigma_X R}\right]^{1/(1-\alpha_N)} = \left[\frac{\lambda_2^*\sigma_N\alpha_N R^{\alpha_N}}{(Ce^{-rt} + \lambda_3^*\sigma_X)R}\right]^{1/(1-\alpha_N)}.$$

Marginal value of on-shop technology. Suppose 'R' is used for on-shop technology development at t. Since the return to scale is α_N, R is converted to R^{α_N} units of on-shop technology. Each unit of on-shop technology has a shadow value of λ_2^*. Due to the uncertainty of development effort, only σ_N of λ_2^* can be an actual contribution provided by a unit of technology, i.e., $\lambda_2^*\sigma_N$. Therefore, the total value obtained by spending R in on-shop technology is $V_n \equiv \lambda_2^*\sigma_N R^{\alpha_N}$. The marginal value of on-shop technology is $\mathrm{d}V_n/\mathrm{d}R = \lambda_2^*\sigma_N\alpha_N R^{\alpha_N-1}$. Thus, the numerator of N^* is merely $\lambda_2^*\sigma_N\alpha_N R^{\alpha_N} = (\mathrm{d}V_n/\mathrm{d}R)R$, the hypothetical total value which would be attained if entire R were spent in developing endogenous technology.

Marginal value of off-shop technology. A similar argument can be made for off-shop technology. Suppose R is allocated for off-shop technology, which is assumed to be converted to R units of off-shop technology (since $\alpha_N \equiv 1$, being normalized). Considering the relative cost advantage of off-shop floor technology over on-shop technology, the total value the firm can attain by devoting R for developing off-shop technology is $V_x \equiv \left(Ce^{-rt} + \lambda_3^*\sigma_X\right)R$.

The marginal value of off-shop technology is $\mathrm{d}V_x/\mathrm{d}R = Ce^{-rt} + \lambda_3^*\sigma_X$. Thus, the denominator of N^* is $\left(Ce^{-rt} + \lambda_3^*\sigma_X\right)R = (\mathrm{d}V_x/\mathrm{d}R)R$, the hypothetical total value which would be attained if entire R were spent in developing exogenous technology. Being scaled by α_N, N^* is determined by the ratio of total marginal values of the two technologies: The larger the total marginal value from on-shop technology, the larger N^* (the more resources the firm needs to allocate for on-shop activities).

Limiting Process ($T \to \infty$)

In this section, we look into the limiting case in order to show a comparable outcome as for the nonlimiting situation.

$$N_{\infty} \equiv \lim_{T\to\infty} N^{*}(t) = \frac{1}{R}\left[\frac{\lambda_{2\infty}\sigma_N\alpha_N}{Ce^{-rt}+\lambda_{3\infty}\sigma_X}\right]^{1/(1-\alpha_N)} = \frac{1}{R}\left[\frac{\beta_N\psi\sigma_N\alpha_N(r+\rho_X)}{C(r+\rho_N)(r+\rho_X)(r+\phi)+\beta_X\psi\sigma_X(r+\rho_N)}\right]^{1/(1-\alpha_N)}. \tag{33}$$

- If C is a constant, N_{∞} remains constant as well, presuming $0 < \frac{\beta_N\psi\sigma_N\alpha_N(r+\rho_X)}{C(r+\rho_N)(r+\rho_X)(r+\phi)+\beta_X\psi\sigma_X(r+\rho_N)} < 1$.
- That $1 < \frac{\beta_N\psi\sigma_N\alpha_N(r+\rho_X)}{C(r+\rho_N)(r+\rho_X)(r+\phi)+\beta_X\psi\sigma_X(r+\rho_N)}$ implies that the 'limiting' value of on-shop technology is larger than that of off-shop technology (indefinitely) so that all the resources are devoted to the on-shop technology development.
- On the other hand, $\frac{\beta_N\psi\sigma_N\alpha_N(r+\rho_X)}{C(r+\rho_N)(r+\rho_X)(r+\phi)+\beta_X\psi\sigma_X(r+\rho_N)} < 0$ implies $C(r+\rho_N)(r+\rho_X)(r+\phi)+\beta_X\psi\sigma_X(r+\rho_N) < 0$, since C is the only term which can be negative. $C(r+\rho_N)(r+\rho_X)(r+\phi)+\beta_X\psi\sigma_X(r+\rho_N) < 0$ can be readily rearranged as $Ce^{-rt}+\lambda_{3\infty}\sigma_X < 0$ (thus, $C_Ne^{-rt}+\lambda_{3\infty}\sigma_X < C_Xe^{-rt}$), and we can derive the same reasoning as that in the nonlimiting case.

2.3.6 Constraint on $R(t)$

We have mainly dealt with a control model without constraining $R(t)$: We have implicitly assumed (i) $R(t)$ is constant within the relevant decision time horizon or (ii) R is reserved only for two activities, on-shop and off-shop technology development, presuming the unused portion of R has no salvage value, and therefore, it is always better to completely consume the entire R between the two mechanisms, i.e., no reason to spare some of R for other uses.

We relax the previous assumptions by replacing them with the new ones: (i) raw resources (i.e., managerial attention) can be used for an activity other than on-shop and off-shop technology development, and we incorporate this feature into the model by allowing a unit of unused resources to have salvage value, η; (ii) the total resources available at a given time is κ, and $R(t)$ is the amount of resources to be allocated for production technology development. Now $R(t)$ is not a constant: Rather, it becomes a dynamic decision variable.

An extended control model encompassing these changes is the second dynamic control problem DCP2:

[DCP2]

$$Maximize \int_0^T (\Psi S(t) - C_N R(t) N(t) - C_X R(t)(1 - N(t)) + \eta(\kappa - R(t)))e^{-rt} dt$$

$$Subject\ to\ \dot{S}(t) = \beta_N A_N(t) + \beta_X A_X(t) - \phi S(t), \tag{34}$$

$$\dot{A}_N(t) = \sigma_N [RN(t)]^{\alpha_N} - \rho_N A_N(t), \tag{35}$$

$$\dot{A}_X(t) = \sigma_X [R - RN(t)]^{\alpha_X} - \rho_X A_X(t), \tag{36}$$

$$0 \le N(t) \le 1, \tag{37}$$

$$0 \le R(t) \le \kappa, \tag{38}$$

$$A_N(0) = A_{N0}, A_X(0) = A_{X0}, S(0) = S_0,$$

$$A_N(T),\ A_X(T),\ \text{and}\ S(T)\ \text{are free.} \tag{39}$$

The current value Hamiltonian of (DCP2) is

$$\begin{aligned} H_C = {} & (\psi S(t) - (C_N - C_X) R(t) N(t) - C_X R(t) + \eta(\kappa - R(t))) \\ & + m_1(\beta_N A_N(t) + \beta_X A_X(t) - \phi S(t)) + m_2(\sigma_N R^{\alpha_N} [N(t)]^{\alpha_N} - \rho_N A_N(t)) \\ & + m_3(\sigma_X R^{\alpha_X} [1 - N(t)]^{\alpha_N} - \rho_X A_X(t)), \end{aligned}$$

where $m_i = \lambda_i e^{rt}$ with λ_i the costate variable for associated differential equation.

The Current value Lagrangian is $\mathcal{L}_C = H_C + n_1(1 - N(t)) + n_2(\kappa - R(t))$, where $n_i = \theta_i e^{rt}$ with θ_i the Lagrangian multiplier for (37) as $i = 1$ and (38) as $i = 2$, respectively.

Since there are two decision variables, $N(t)$ and $R(t)$, the optimality condition requires two first-order conditions to be satisfied simultaneously:

$$\begin{aligned} \frac{\delta \mathcal{L}_C}{\delta N} & = -(C_N - C_X)R + m_2 \sigma_N \alpha_N R^{\alpha_N} [N(t)]^{\alpha_N - 1} - m_3 \sigma_X \alpha_X R^{\alpha_X} [1 - N(t)]^{\alpha_X - 1} - n_1 \\ & = 0 \end{aligned}$$

and

$$\begin{aligned} \frac{\delta \mathcal{L}_C}{\delta R} & = -(C_N - C_X)N - C_X \\ & \quad - \eta + m_2 \sigma_N \alpha_N R^{\alpha_N - 1} N^{\alpha_N} + m_3 \sigma_X \alpha_X R^{\alpha_X - 1} [1 - N(t)]^{\alpha_X} - n_2 \\ & = 0, \end{aligned}$$

$$\frac{\delta \mathcal{L}_C}{\delta n_1} \geq 0, n_1^*(t) \geq 0, n_1(t)^*[1 - N^*(t)] = 0,$$
$$\frac{\delta \mathcal{L}_C}{\delta n_2} \geq 0, n_2^*(t) \geq 0, n_2(t)^*[\kappa - R^*(t)] = 0.$$

The sufficient condition calls for

$$\frac{\delta^2 \mathcal{L}_C}{\delta N^2} = m_2\sigma_N\alpha_N(\alpha_N - 1)R^{\alpha_N}[N(t)]^{\alpha_N - 2} + m_3\sigma_X\alpha_X(\alpha_X - 1)R^{\alpha_X}[1 - N(t)]^{\alpha_X - 2} \leq 0,$$
$$\frac{\delta^2 \mathcal{L}_C}{\delta R^2} = m_2\sigma_N\alpha_N(\alpha_N - 1)R^{\alpha_N - 2}N^{\alpha_N} + m_3\sigma_X\alpha_X(\alpha_X - 1)R^{\alpha_X - 2}[1 - N]^{\alpha_X} \leq 0.$$

There is no change in the costate variables. The sufficient condition associated with R can be proved in virtually the same way as in Observation 1. As in [DCP1], we further assume $\alpha_X = 1$ through some normalization process.

$$N^*(t) = \left[\frac{m_2\sigma_N\alpha_N R^{\alpha_N}}{(C_N - C_X)R + m_3\sigma_X R + n_1}\right]^{1/(1-\alpha_N)} \tag{40}$$

$$R^*(t) = \left[\frac{m_2\sigma_N\alpha_N N^{\alpha_N}}{((C_N - C_X) + m_3\sigma_X)N + C_X + \eta - m_3\sigma_X + n_2}\right]^{1/(1-\alpha_N)} \tag{41}$$

Theorem 1 (Shadow price of managerial attention). *Let us focus on $t \in \{t | n_1 = 0 \, and \, t \in [0, T]\}$, i.e., the time period in which $N^* < 1$. The shadow price of managerial attention is as follows: $\theta_2 = n_2 e^{-rt} = \lambda_3^* \sigma_X - C_X e^{-rt} - \eta e^{-rt}$. Since $\theta_2 \geq 0$, in order for [DCP2] to have a feasible solution $\lambda_3^* \sigma_X - C_X e^{-rt} \geq \eta e^{-rt}$ should be valid. This condition implies that if the current value benefit from the alternative use of the resources (ηe^{-rt}) is larger than the net benefit from off-shop technology ($\lambda_3^* \sigma_X - C_X e^{-rt}$), the firm should not allocate resources for production technology development through either on-shop or off-shop mechanism. The threshold value is determined with the off-shop technology net value.*

Proof With $n_1 = 0$, $N^*(t) = \frac{1}{R}\left[\frac{m_2\sigma_N\alpha_N}{C + m_3\sigma_X}\right]^{1/(1-\alpha_N)}$.

From (41),

$$\begin{aligned}(N^*R^*)^{1-\alpha_N} &= N^{*1-\alpha_N}\frac{m_2\sigma_N\alpha_N N^{\alpha_N}}{((C_N - C_X) + m_3\sigma_X)N + C_X + \eta - m_3\sigma_X + n_2} \\ &= \frac{m_2\sigma_N\alpha_N N}{(C + m_3\sigma_X)N + C_X + \eta - m_3\sigma_X + n_2} = \frac{m_2\sigma_N\alpha_N}{C + m_3\sigma_X}\end{aligned} \tag{42}$$

Therefore, from (42), $N^*(C + m_3\sigma_X) = (C + m_3\sigma_X)N^* + C_X + \eta - m_3\sigma_X + n_2$, and thus $n_2 = m_3\sigma_X - C_X - \eta \Rightarrow \theta_2 = n_2 e^{-rt} = \lambda_3^*\sigma_X - C_X e^{-rt} - \eta e^{-rt}$.

Thus, the theorem is proved. The theorem implies that if $\lambda_3^* \sigma_X - C_X e^{-rt} = \eta e^{-rt}$, then $R^*(t) < k$: In other words, if the profit from the alternative use of the resources is equal to the net value of off-shop technology, then the firm does not have to use the entire resources for production technology development (some of the resources might be diverted to the alternative activity). Theorem 1 offers a boundary condition on the shadow value of resources (managerial attention) and makes the constraints of control model tighter.

2.3.7 Manufacturing Learning-Induced Model

In order to be consistent with the manufacturing learning theory, we need to consider t as the cumulative production units: [0, T] still has the meaning of 'relevant decision time horizon' expressed in terms of the production unit rather than time unit.

A widely accepted formulation of learning phenomenon is suggested (Yelle 1979):

$$Y = KX^n, \tag{43}$$

where Y is the number of direct labor hours required to produce the Xth unit, K the number of direct labor hours required to produce the first unit, X the cumulative unit number, $n = \log\Xi/\log 2$ the learning index, with Ξ the learning rate, and $1 - \Xi$ the progress rate. (43) is designed so that whenever the number of cumulative units doubles, Y is reduced by $(1 - \Xi)\%$: For instance, if $K = 10$ and $\Xi = 0.8$, then $Y = 8$ when $X = 2$.

Assuming that the managerial learning propensity is associated with the implementation cost, the learning-induced control model incorporates the learning factor into the opportunity cost structure consisting of C_N and C_X, $C_{L_i} \equiv C_i t^{b_i}$ where $i = N$ or X, t is the cumulative production unit, $b_i = \frac{\log \Xi_i}{\log 2}$ the learning index with Ξ_i the learning rate for on-shop mechanism ($i = N$) or off-shot mechanism ($i = X$). Thus, C_{L_i} is the learning-induced cost of on-shop ($i = N$) or off-shop ($i = X$) technology.

The dynamic (learning-induced) opportunity cost structure becomes

$$C(t) = C(t)_{L_N} - C(t)_{L_X} = C_N t^{b_N} - C_X t^{b_X}. \tag{44}$$

One of the initial model assumptions is $C_N < C_X$, i.e., without learning-induced evolution, the static cost structure consists of C_N and C_X such that $C_N < C_X$, which we assume in general (saying the implementation cost associated with off-shop technology is more expensive than that with on-shop technology).

Therefore, we know that $C(0) < 0$. From (44), we can evaluate τ such that from τ on $C(t) \geq 0$, i.e., $C(t)_{L_N} \geq C(t)_{L_X}$. As we have seen in the nonlearning control model, the sign of $C(t)$ determines the overall dynamics of technology development between on-shop and off-shop mechanisms. If we incorporate the learning factor into the cost structure, the sign of $C(t)$ might change over time: In particular, one important change is from $C(t) < 0$ to $C(t) \geq 0$. This is a critical shift since it implies that the cost advantage moves from on-shop technology to off-shop technology, and thus, the firm may have to experience a significant redirection of technological development.

The shifting point of time, τ, can be assessed: Assuming $\Xi_N > \Xi_X$,

$$\begin{aligned} & C_N t^{b_N} - C_X t^{b_X} \geq 0 \\ & \Rightarrow \frac{C_N}{C_X} \geq t^{b_X - b_N} \\ & \Rightarrow t \geq (C_X/C_N)^{1/b_N - b_X} = (C_X/C_N)^{\log 2/(\log(\Xi_N/\Xi_X))}. \end{aligned} \tag{45}$$

Thus, $\tau = (C_X/C_N)^{\log 2/(\log(\Xi_N/\Xi_X))}$, at which the sign of $C(t)$ changes and it becomes possible for N^* to decrease dramatically over time (after τ). However, if $(C_X/C_N)^{\log 2/(\log(\Xi_N/\Xi_X))} \geq T$, then $C(t) < 0$ throughout [0,*T*], the relevant decision time horizon. Thus, in order for $C(t)$ to become a positive number given the current decision time span, the learning rate difference between on-shop and off-shop technologies should be at least large enough to ensure, assuming $T > 1$,

$$\begin{aligned} & (C_X/C_N)^{\log 2/(\log(\Xi_N/\Xi_X))} < T \\ & \Rightarrow (\ln 2/(\ln(\Xi_N/\Xi_X)) \ln(C_X/C_N) < \ln T \\ & \Rightarrow \exp\{(\ln 2 \ln(C_X/C_N))/\ln T\} < \Xi_N/\Xi_X. \end{aligned}$$

We have proved Theorem 2.

Theorem 2 (Learning-induced cost structure). *Given the specified dynamic cost structure above, $C(t)$ changes its sign from negative to positive at τ such that $\tau = (C_X/C_N)^{\log 2/(\log(\Xi_N/\Xi_X))}$. If $\exp\{(\ln 2 \ \ln(C_X/C_N))/\ln T\} < \Xi_N/\Xi_X$, we can expect N^* to experience an accelerating rate to decrease within the current decision horizon, [0, T]: that is to say, the firm needs to increase the proportion of resources to be allocated for off-shop technology development (sufficient condition for $N^*(T) = 0$).*

Example 1 Figure 5 shows the two dynamics of cost structure, one without and the other with learning-induced evolution. It is based on the parameters such that $C_X = 5, C_N = 3, \Xi_X = 0.75, \Xi_N = 0.95$.

Theorem 2 gives us the exact timing of shift, $\tau = 4.4721$ (cumulative) units, which is consistent with the figure.

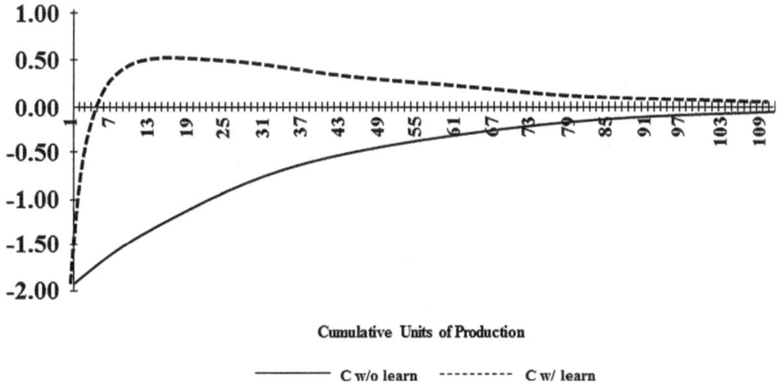

Fig. 5 Cost structure with learning effect

2.3.8 Dynamic Learning Model

The previous section of 'value interpretation' has dealt with the static learning model, assuming a fixed value of C within a decision time horizon. In effect, the static model implicitly presumes that the learning happens in a discrete way so that the firm experiences one fixed implementation cost structure (possibly, induced by learning) for a given cycle of technology development. Although the static premise seems compatible with the proposition regarding manufacturing learning capability, it is not suitable for describing a situation in which the decision time horizon is long enough to encompass the evolving effect of manufacturing learning: It is the dynamic learning model which can complement the static model in this regard.

The dynamic learning model proves to be able to duplicate the resource or attention allocation patterns derived by the static model: We can have three broad categories of resource allocation dynamics, (i) on-shop technology-induced learning in Fig. 6 (on-shop learning rate = 0.7; off-shop learning rate = 1), (ii) neutral

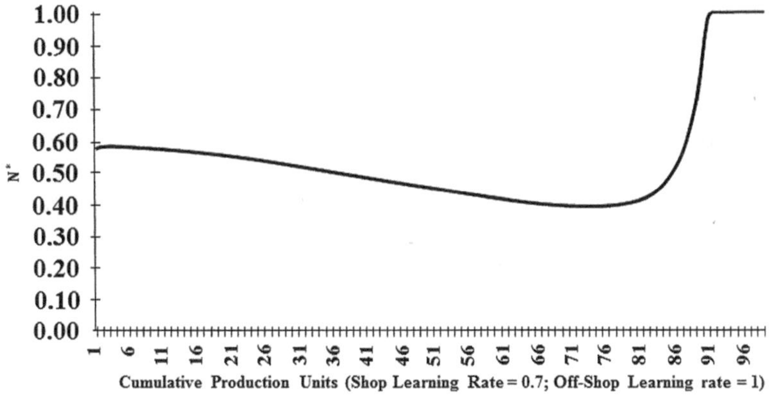

Fig. 6 Shop floor technology development effort

learning in Fig. 7 (on-shop learning rate = 1; off-shop learning rate = 0.83), (iii) off-shop technology-induced learning in Fig. 8 (on-shop learning rate = 1; off-shop learning rate = 0.7) based on the parameter values given in Table 3.

The dynamic model specifically assumes that the learning affects the implementation cost as the experience accumulates (i.e., the number of production units increases): It is a consistent postulate with the manufacturing learning theory.

As shown in Fig. 5, the big chunk of learning realizes in the early period of decision horizon, but the main impact of manufacturing learning becomes apparent only after almost a half (in general) of the decision time span passes. Thus, what is critical in understanding the overall dynamics is to investigate fully the momentous changes in the direction of the dynamics in the later period: The analysis clearly indicates that the management learning intention or propensity biased toward a particular technological advancement (i.e., either on-shop technology or off-shop

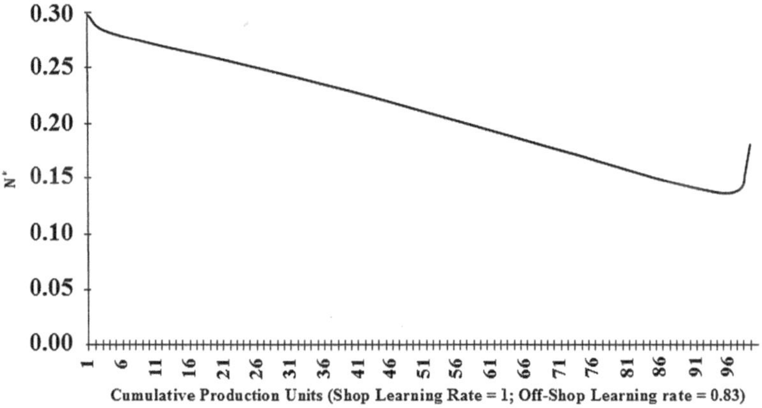

Fig. 7 Shop floor technology development effort

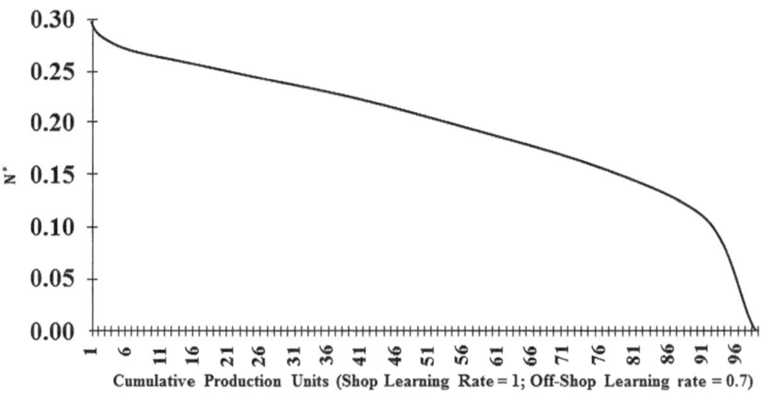

Fig. 8 Shop floor technology development effort

Table 3 Parameter values in Figs. 6, 7 and 8

ψ	ϕ	r	β_N	β_X	ρ_N	ρ_X	α_N	σ_N	σ_X	R	T	$C_N - C_X$
1	0.03	0.03	0.7	1.0	0.009	0.019	0.75	0.9	0.8	1	100	2 − 7 = −5

technology) largely determines the time point as well as the rate to accelerate the move toward a direction consistent with the specific development mechanism in the later phase.

2.4 Inferences and Implications

A key determinant for production technology development is the managerial attention from managers participating in the development activities. In fact, the managerial attention is the most precious resources in a firm (March and Olsen 1976). We propose that the *manufacturing learning propensity espoused by (middle) managers* toward a particular technological direction is a significant factor to shape the attention allocation dynamics.

In the relatively early stage of decision time horizon, the management forms a learning propensity for either on-shop technology or off-shop technology, and endeavors to make decisions and behave according to the formed propensity. Since in the early period the benefit from technological improvement far outweighs the implementation cost, the learning impact seems minimal. However, as the terminal point of decision horizon approaches, the implementation cost becomes dominant over the benefit (since the benefit decreases more rapidly than the cost does). The firm may eventually face a time point at which the rate to move toward the direction conformable with the 'selected' technology accelerates so enormously that the firm cannot control its own dynamics with just a minor adjustment: The dynamics of the later period was already prophesied by the initial choice of learning propensity (*freezing effect of learning intention*; *catastrophic effect of learning prophecy*).

The result underlines the *criticality of learning propensity developed by the management* for a certain type of production technology in determining the optimal dynamics of technology development: Small decisions accumulated by complying with a benign choice of particular learning propensity at the early stage might become an irreversible force the firm could not deny to follow in the later stage (Figs. 9, 10, and 11). Thus, top management must be able to understand the early formation of *learning propensity/intention by (middle) management* in order to optimally manage the dynamics of technology development.

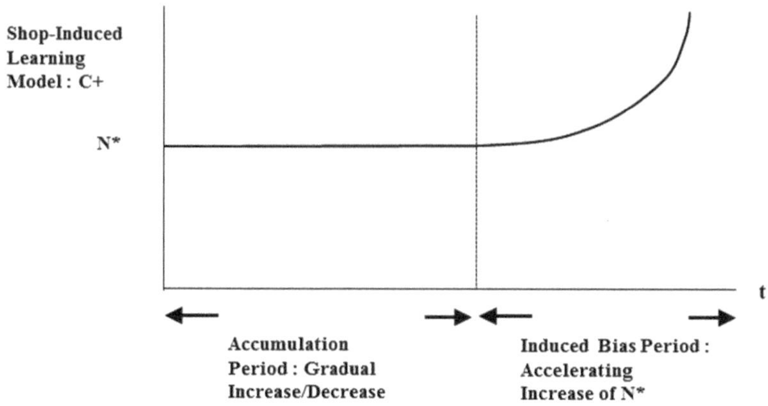

Fig. 9 On-shop-induced learning model (C+)

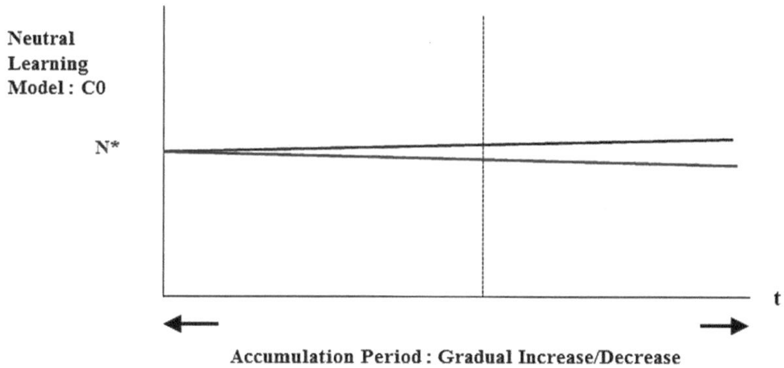

Fig. 10 Neutral learning model (C0)

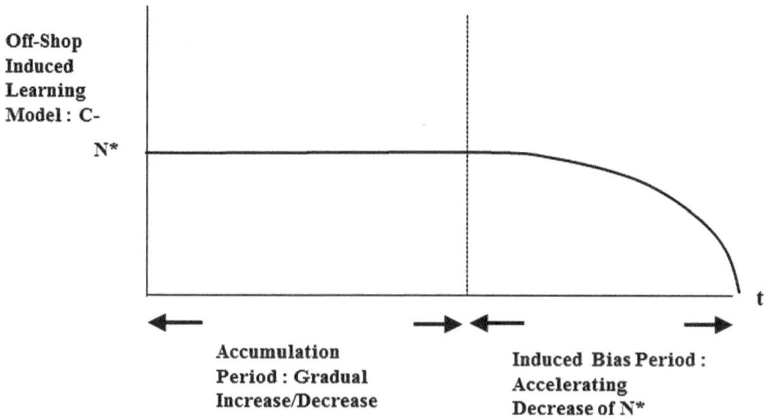

Fig. 11 Off-shop-induced learning model (C−)

Exercise Problems

1. Find necessary conditions for the solution of the following problem:

$$Max_u \int_0^7 \left[4x - 2u - u^2\right] \mathrm{d}t$$
$$Subject\ to\ \dot{x}(t) = x + 2u,\ x(0) = 3$$

2. Show that the necessary conditions above are also sufficient for optimality.
3. An optimal control problem is given as follows:

$$Max_u \int_0^2 [2x - 4u] \mathrm{d}t$$
$$Subject\ to\ \dot{x}(t) = x + u, x(0) = 2,\ 0 \le u \le 1$$

   (a) Find the optimal control $u(t)$ that solves the problem.
   (b) Interpret the result in part (a).

4. Solve the minimization problem:

$$Min_u \int_0^2 \left[x + u^2\right] \mathrm{d}t$$
$$Subject\ to\ \dot{x}(t) = x + u, x(0) = 1$$

5. An optimal control problem is given as follows:

$$Max \int_0^3 (x_1 + x_2 - 4u^2 - v^2) \mathrm{d}t + 2x_1(3)$$
$$Subject\ to\quad \dot{x}_1(t) = x_1 + u, x_1(0) = 1$$
$$\dot{x}_2(t) = v(t), x_2(0) = 0$$

   (a) Suggest necessary conditions.
   (b) Characterize $u(t)$ and $v(t)$ that solve the problem and compare their dynamics.

6. Find the optimal path $u(t)$ of the following optimal control problem.

$$Max \int_0^1 (2 + 4t) u \mathrm{d}t$$
$$Subject\ to\quad \dot{x}(t) = 2x + 4tue^{2t}, x(0) = 0, x(1) = e^2$$
$$-1 \le u(t) \le 1$$

7. A monopolist needs to make a decision on production and investment in process knowledge in order to maximize its profit over the planning horizon $[0, T]$. Let $q(t)$ be the quantity the firm produces and sells at time t. Market price is determined by $p(q) = a - bq$. Let $k(t)$ denote the stock of process knowledge of the firm at t and $u(t)$ denote the knowledge investment rate at t. Process knowledge $k(t)$ increases by two types of learning: (1) autonomous learning that results from the cumulative production experience and (2) induced learning that results from the cumulative knowledge investment effort such as process improvement projects and quality circles. Specifically, the dynamic equation of process knowledge accumulation is $\dot{k}(t) = \alpha q(t) + \beta u(t)$, where $\alpha \geq 0$ and $\beta \geq 0$ represent the relative contribution of autonomous and induced learning to the firm's process knowledge. Firm's process knowledge stock $k(t)$ determines the unit production cost $c_1(k)$, i.e., $c_1(k)$ is decreasing in k and $\lim_{k\to\infty} c_1(k) > 0$. Also, the cost of knowledge investment is given by $c_2(u)$, where $c_2(u)$ is an increasing convex function of $u(t)$. Let r be the discounting rate and $k(0) = k_0$
   (a) Formulate the current value Hamiltonian for the optimal control model of the problem.
   (b) Derive the necessary conditions for the optimal solution.
8. A manufacturing firm produces and sells its product at a rate q at time t. Market price p is inversely related to q, i.e., $p'(q) < 0$. But the firm experiences a learning curve effect; thus, the unit production cost c decreases with the cumulative production volume Q, i.e., $c = c(Q), c'(Q) < 0$. Assume that the firm wants to determine the optimal production rate $q(t)$ to maximize its profit over $t \in [0, T]$. Then, the optimal control model of the problem is as follows:

$$Max \int_0^T [(p(q) - c(Q))q]dt$$

$$Subject\ to\ \dot{Q}(t) = q, Q(0) = 0$$

   (a) Derive the necessary conditions for the optimal solution.
   (b) Characterize the optimal production rate $q(t)$.

Appendix 1: Derivation of Eqs. (27)–(29)

$$\dot{m}_1 = -\frac{\delta \mathcal{L}_c}{\delta S} + rm_1 = (r + \phi)m_1 - \psi, \tag{46}$$

therefore,

$$m_1 = K_1 e^{(r+\phi)t} + \frac{\psi}{r+\phi}, \; K_1 = \text{constant}. \tag{47}$$

From $m_1(T) = 0$ along with (47), we can get

$$K_1 = -\frac{\psi}{r+\phi} e^{-(r+\phi)T} \tag{48}$$

From (47) and (48) with $m_1 = \lambda_1 e^{rt}$, (27) follows:

$$\dot{m}_2 = -\frac{\delta \mathcal{L}_c}{\delta A_N} + rm_2 = (r+\rho_N)m_2 - \beta_N m_1. \tag{49}$$

Let

$$m_2 = K_2 e^{(r+\rho_N)t} + L_2 e^{(r+\phi)t} + M_2, \tag{50}$$

where K_2, L_2, and M_2 are constants. Then, from (50),

$$\dot{m}_2 = K_2(r+\rho_N)e^{(r+\rho_N)t} + L_2(r+\phi)e^{(r+\phi)t}. \tag{51}$$

Substituting (47) into (49),

$$\dot{m}_2 = (r+\rho_N)m_2 - \beta_N K_1 e^{(r+\phi)t} - \frac{\beta_N \psi}{r+\phi} \tag{52}$$

From (50) and (51),

$$\dot{m}_2 - (r+\rho_N)m_2 = L_2(r+\phi)e^{(r+\phi)t} - L_2(r+\rho_N)e^{(r+\phi)t} - (r+\rho_N)M_2. \tag{53}$$

From (52),

$$\dot{m}_2 - (r+\rho_N)m_2 = -\beta_N K_1 e^{(r+\phi)t} - \frac{\beta_N \psi}{r+\phi}. \tag{54}$$

By comparing (53) and (54),

$$L_2 = \frac{-\beta_N K_1}{\phi - \rho_N} \text{ and } M_2 = \frac{\beta_N \psi}{(r+\phi)(r+\rho_N)}. \tag{55}$$

By plugging (55) and (48) into (50) and applying $\mathrm{m}_2(T) = 0$,

$$K_2 = -\frac{\beta_N \psi}{(\phi - \rho_N)(r + \rho_N)} e^{-(r+\rho_N)T}. \tag{56}$$

Since $m_2 = \lambda_2 e^{rt}$, (28) follows. Adopting similar steps, we can derive (29).

Appendix 2: Basics of Differential Equations

In this appendix, we review optimization models and discuss basic rules of differential equations. Our objective is not to teach the student complete theories for these subjects, but to present key features in an integrated way to refresh the student's prior knowledge about these subjects. If the student does not have any background on some of these subjects or finds it necessary to learn them almost from scratch, then he should thoroughly study the textbooks that cover these subjects.

Types of Optimization Models

There are different types of optimization problems, depending on time and dynamic nature. Table 4 presents the typology of optimization models, using the two dimensions, time and dynamic nature. For the time dimension, there are two modes, discrete time and continuous time. Regarding the dimension of dynamic nature, there are two models, nondynamic model and dynamic model. A dynamic model is the one, where the previous period's state affects the current period's. Therefore, a dynamic model is also a multi-period model.

We discuss the optimization models in Table 4.

Table 4 Typology of optimization models

Optimization modeling		Time	
		Discrete	Continuous
Nondynamic	Single-period	$Max_{y\geq 0} G(y)$	
	Multi-period	$Max \sum_{t=1}^{T} G(t, y_t)$ $Subject\ to\ y_t \geq 0, t = 1, \ldots, T$	$Max \int_0^T G(t, y(t))dt$ $Subject\ to\ y(t) \geq 0$
Dynamic and multi-period		$Max \sum_{t=1}^{T} G(t, y_t, y_{t-1})$ $Subject\ to\ y_t \geq 0, t = 1, \ldots, T$ y_0 given	$Max \int_0^T G(t, y(t), y'(t))dt$ $Subject\ to\ y(t) \geq 0$ $y(0) = y_0$ given

Discrete-Time Models

A discrete-time model is the one, where the time period is identified discretely, e.g., period 1 and period 2.

Nondynamic Models

A single-period model is as follows:

$Max_{y \geq 0} G(y)$, assuming a nonnegative decision variable, y, i.e., $y \geq 0$.

On the other hand, a multi-period model is formulated as follows:

$$Max \sum_{t=1}^{T} G(t, y_t)$$
$$Subject\ to\ y_t \geq 0,\ \ \mathrm{t} = 1, \ldots, \mathrm{T}$$

That is, we optimize the objective function over T discrete periods.

Dynamic Models

As alluded to before, a dynamic model is the one, where the previous period's state affects the current period's state. An example of a discrete-time dynamic model is as follows:

$$Max \sum_{t=1}^{T} G(t, y_t, y_{t-1})$$
$$Subject\ to\ y_t \geq 0,\ \ \mathrm{t} = 1, \ldots, \mathrm{T},\ \ \text{and}\ y_0\ \text{given.}$$

Continuous-Time Models

In a continuous-time model, we integrate the objective function over the continuous time.

Nondynamic Models

An example of a continuous-time nondynamic model is given as follows:

$$Max \int_0^T G(t, y(t))dt$$
$$Subject\ to\ y(t) \geq 0$$

Dynamic Models

A continuous-time dynamic model is as follows:

$$Max \int_0^T G(t, y(t), y'(t))dt$$
$$Subject\ to\ y(t) \geq 0, y(0) = y_0 \text{ given.}$$

Note that $y'(t) = \frac{dy(t)}{d_t}$ is equivalent to y_{t-1} in the discrete-time model.

Simple Differential Equations

In this section, we present basic differential equations, which are useful to solve an optimal control theory problem.

Basic Principle

If we integrate $x(t) = t + 1$, we have $X(t) = \int x(t)dt = \frac{1}{2}t^2 + t + c$, where c is a constant to be determined. As such, if we differentiate $X(t) = \frac{1}{2}t^2 + t + c$ with regard to t, we get $x(t) = t + 1$.

$$x(t) = t + 1 \underset{\overleftarrow{differentiate}}{\overset{\overrightarrow{integrate}}{\rule{2cm}{0.4pt}}} X(t) = \int x(t)dt = \frac{1}{2}t^2 + t + c, \text{ where } c \text{ is a constant.}$$

Similarly, if we integrate $x(t) = -ke^{-kt}$, we get $X(t) = \int x(t)dt = e^{-kt} + c$, where c is a constant to be determined. In turn, if we differentiate $X(t) = \int x(t)dt = e^{-kt} + c$, we have $x(t) = -ke^{-kt}$.

$$x(t) = -ke^{-kt} \underset{\overleftarrow{differentiate}}{\overset{\overrightarrow{integrate}}{\rule{2cm}{0.4pt}}} X(t) = \int x(t)dt = e^{-kt} + c, \text{ where } c \text{ is a constant.}$$

Now suppose we have a differential equation $x(t)' + x(t) = c$, where c is a constant. Then how can we get $x(t)$?

Let us try $x(t) = e^{-t} + c$, where c is a constant. Then, $x(t)' = -e^{-t}$ and therefore, $x(t)' + x(t) = c$.

Consider another differential equation $x(t)' + kx(t) = h$, where k, h are constants.

Let us try $x(t) = e^{-kt} + c$, where c is a constant. Then, $x(t)' = -ke^{-kt}$, and therefore, $x(t)' + kx(t) = kc = h$, $\therefore c = \frac{h}{k}$. Finally, $\therefore x(t) = e^{-kt} + \frac{h}{k}$

Integrating Factor

In the above section, we have relied on a sort of trial-and-error approach. There is a more formal method, which uses the integrating factor.

Consider a differential equation $x(t)' + kx(t) = c$, where c, k are constants. Since $\frac{d(e^{kt}x(t))}{dt} = ke^{kt}x(t) + e^{kt}x(t)' = e^{kt}(x(t)' + kx(t)) = e^{kt}c$, where e^{kt} is the integrating factor, $\frac{d(e^{kt}x(t))}{dt} = e^{kt}c$, and thus, $e^{kt}x(t) = \int ce^{kt}dt$. In addition, $e^{kt}x(t) = \int ce^{kt}dt = \frac{c}{k}e^{kt} + h$, where h is a constant, $\therefore x(t) = \frac{c}{k} + he^{-kt}$.

Using $x(0) = x_0$ given initial condition, we obtain $x(0) = \frac{c}{k} + h = x_0$; thus, $h = x_0 - \frac{c}{k}$.

$$\therefore x(t) = \frac{c}{k} + he^{-kt} = \frac{c}{k} + \left(x_0 - \frac{c}{k}\right)e^{-kt}$$

Appendix 3: Current Value Hamiltonian

Consider an optimal control theory problem with a discounting factor r:

$$\begin{aligned} &Max \int_{t_0}^{t_1} e^{-rt} f(t, x(t), u(t))dt \\ &Subject\ to \quad x'(t) = g(t, x(t), u(t)) \\ &\qquad\qquad\quad x(t_0) = x_0, t_0, t_1 \text{ fixed (given), and } x(t_1) \text{ free (unfixed)} \end{aligned}$$

To solve the problem, we can use the Pontryagin's maximum principle. As usual, we first form the Hamiltonian: $H = e^{-rt}f(t, x(t), u(t)) + \lambda(t)g(t, x(t), u(t))$.

Although it is possible to apply the regular maximum principle, we suggest a simpler or more convenient form, i.e., the current value Hamiltonian $\mathbf{H}$:

$$\mathbf{H} \equiv e^{rt}H = f(t, x(t), u(t)) + m(t)g(t, x(t), u(t)), \text{where } m(t) \equiv e^{rt}\lambda(t).$$

Then, the necessary conditions become as follows:

(1) $\partial\mathbf{H}/\partial u = f_u + mg_u = 0,$
(2) $m' = rm - \partial\mathbf{H}/\partial x = rm - f_x - mg_x,$
(3) $\lambda(t_1) = e^{-rt_1}m(t_1) = 0$ if $x(t_1)$ free,
(4) $\lambda(t_1) = e^{-rt_1}m(t_1) \geq 0$ and $e^{-rt_1}m(t_1)x(t_1) = 0$, if $x(t_1) \geq 0,$
(5) $x' = \partial\mathbf{H}/\partial m = g,$
(6) $x(t_0) = x_0.$

Appendix 4: Bounded Controls

Let us consider an optimal control theory problem with a bounded control variable:

$$Max \int_{t_0}^{t_1} f(t, x(t), u(t))\mathrm{d}t$$

$$\begin{aligned} \textit{Subject to} \quad & x'(t) = g(t, x(t), u(t)), x(t_0) = x_0 \\ & b_l \leq u(t) \leq b_u. \end{aligned}$$

To solve the problem, we first need to form the Hamiltonian, H:

$$H(t, x(t), u(t), \lambda(t)) = f(t, x(t), u(t)) + \lambda(t)g(t, x(t), u(t)).$$

In addition, we form the Lagrangian with multipliers μ_1, μ_2:

$$\begin{aligned} L &= H + \mu_1(b_u - u) + \mu_2(u - b_l) \\ &= f(t, x, u) + \lambda g(t, x, u) + \mu_1(b_u - u) + \mu_2(u - b_l). \end{aligned}$$

Then, the necessary conditions are as follows:

$$\frac{\partial L}{\partial u} = f_u + \lambda g_u - \mu_1 + \mu_2 = 0,$$

$$\mu_1 \geq 0, \ \mu_1(b_u - u) = 0,$$

$$\mu_2 \geq 0, \ \mu_2(u - b_l) = 0,$$

and plus other necessary conditions from the maximum principle such as $\lambda' = -\frac{\partial H}{\partial x}$ and appropriate transversality conditions.

Chapter 2
Value of Coordination in Supply Chain Management

Abstract The most important subject in operations strategy is supply chain management, which in turn emphasizes coordination as the infrastructural dimension. In effect, coordination is an essential key to effective supply chain management. Against this backdrop, in this chapter, I discourse on how to define the value of coordination in supply chain management.

Keywords Supply chain management · Coordination

Key Learning

- *What is the value of coordination in supply chain management?*
  - Coordination between supply chain partners is critical to effective supply chain management.
  - The balanced decision-making forges a better outcome than the dominated case does.
  - The value of the balanced decision-making comes from more effective resource utilization than the dominated case does.
- *How should the supply chain partners coordinate with each other for innovation?*
  - We consider a particular supply chain situation in which the manufacturer coordinates with its supplier for the supplier's innovation that can eventually lead to supply cost reduction.
  - Although the coordination could improve the manufacturing firm's own profitability, it might not be attractive to the supplier unless the supply cost reduction should ultimately increase the market demand to a certain extent.

The original version of this chapter was revised. The erratum to this chapter is available at 10.1007/978-981-10-3599-9_6

B. Kim, *Optimal Control Applications for Operations Strategy*,
DOI 10.1007/978-981-10-3599-9_2

In this chapter, we would like to understand the dynamic nature of supply chain coordination. We first are concerned with the effective structure of decision-making in supply chain management and consider two cases of decision-making structure for a simple supply chain consisting of two players: the first case in which one supply chain partner dominates the entire decision-making process, and the other in which two players share the decision-making process equally. Since the literature in general supports that the balanced decision-making forges a better performance than the dominated case does, our primary objective is to analyze detailed dynamics of resource allocation and sources of the benefit of the balanced decision-making. It turns out that the value of the balanced decision-making comes from more effective resource utilization than the dominated case does, i.e., each of the cooperative partners knows how to optimize its resource utilization better than the other does.

Then, we delve into a more specific question in supply chain management, i.e., whether it is always better for the supply chain partners to coordinate with each other or under what circumstances it is more desirable to coordinate. In order to sustain supply chain partnership, coordination should enhance the profitability of not only the manufacturer, but also the supplier(s). We consider a setting, where the supply chain partners should decide to coordinate with each other for innovation, i.e., the manufacturer should decide whether to coordinate with its supplier for the supplier's innovation that can eventually lead to supply cost reduction. We show that although the coordination could improve the manufacturing firm's own profitability, it might not be attractive to the supplier unless the supply cost reduction should ultimately increase the market demand to a certain extent. Under particular circumstances, e.g., when the market demand stays constant, the manufacturer's profit increase due to the coordination equals the amount of profit loss to the supplier.

1 Joint Decision-Making in Supply Chain Management[1]

For effective supply chain management, coordination between supply chain partners is critical. But, there is a subtle issue: Depending on the bargaining power balance between the supply chain participants, it is determined who exerts more influence in making decisions related with such coordination. For example, it is often possible that one company in a chain might attempt to influence other members in order to achieve its own goals and promote its own interests.

[1]This section is a refined version of Kim, B. and J. Kim (2008) "Values of Balanced Decision-Making between Supply Chain Partners." *International Transactions in Operational Research, 15*, 623–649.

1.1 Decision-Making Structure

Let us consider two cases of decision-making structure in the context of a simple supply chain consisting of two players: the first case in which a supply chain partner dominates the decision-making process and the other partner passively follows the dominant player's decision, and the other case in which the two players share the decision-making process equally in a balanced way. For instance, in an automobile industry, a supplier that does not have a strong bargaining power cannot help but act passively vis-à-vis its stronger partner, e.g., the carmaker.

In the game theory literature, it is well established that a collusive or cooperative case generates a better result than a competitive case from the entire system's perspective: This result is mainly due to the existence of free-riding. Applying the reasoning to the current context, we know a priori that the balanced decision-making forges a better outcome than the dominated case does for the supply chain system as a whole. In this section, therefore, rather than asking about which decision-making structure is better for the supply chain, our primary questions are '*How do the resource allocation dynamics between the supply chain partners behave?*' and '*Where comes the benefit of the balanced case?*' To answer the questions, we set up optimal control theory models and derive analytical solutions, which indicate that the value of the balanced decision-making comes from more effective resource utilization than the dominated case does.

Although a game-theoretic approach has a few advantages, one of which is to enable the researcher to analyze competitive interactions between players (e.g., supply chain partners) rigorously, it also has a serious disadvantage, i.e., its inability to analyze the competitive situation dynamically over a longer period time in a continuous time line. For instance, while a game-theoretic model can only suggest a conclusion that the total profit under a centralized decision-making structure is larger than that under a decentralized structure eventually, an optimal control theory model can show a dynamic pattern of the total profit, e.g., the centralized structure's total profit lags behind the decentralized structure's profit for the initial period, but it surpasses the other since the midpoint of the decision time horizon: If a usual game-theoretic model is for snapshot displays, an optimal control theory model is for movies. That is, with the optimal control theory model, we are more capable of analyzing the optimal

Fig. 1 Comparison of research methodologies

	Static	Dynamic
Competitive	Usual Game - Theoretic Model	Optimal Control Theory Model
Non-competitive	Usual Empirical Case Studies	Simulation based on Empirical Cases

dynamics and understanding intricate changes of the fundamental forces that drive such dynamics. Figure 1 compares different research methodologies.

1.2 Optimal Control Theory Models

There are many areas of supply chain coordination. Supply chain partners can coordinate for sharing information about market demand, improving quality of an existing product, conducting a joint marketing or promotion, making joint decisions on plant capacity or inventory level, and the like. In this section, we focus on a specific area of coordination, *new product development.*

We consider a simple supply chain consisting of two players, a supplier and a manufacturer. To highlight the difference in decision-making structures, two cases are studied. The first case portrays a situation where the manufacturer dominates the decision-making process. The manufacturer's domination is embodied in its ability to make use of the entire resources in the supply chain including the supplier's as well as its own. In the second case, on the other hand, the supplier is also actively involved in the decision-making process by participating in the decision-making for resource allocation.

1.2.1 A Case in the Automotive Industry

In the automobile industry, there usually exists a huge unbalance in bargaining power between the supply chain partners, in particular a carmaker (manufacturer) and its supplier(s): In general, it is the manufacturer that has the larger bargaining power. Sometimes, the player with a dominating bargaining power wields its excessive influence on the weaker player to gain its own profit at the cost of the other. For instance, there is a quote from a director of a major supplier to the big three US carmakers,[2] saying 'The Big Three (U.S. automakers) set annual cost-reduction targets (for the parts they purchase). To realize those targets, they'll do anything. (They've unleashed) a reign of terror and it gets worse every year.' This is a situation comparable to the manufacturer-dominating decision-making structure mentioned above, where the manufacturer exerts its excessive bargaining power to coerce the weaker supplier into passively accepting the terms and conditions without any meaningful bilateral negotiations. Therefore, the manufacturer obtains a de facto right to utilize the supplier's resources at its own '*arbitrary*' discretion. The question is, '*Is this type of manufacturer-dominating decision structure in a supply chain good even for the manufacturer itself?*' But, this kind of coercive relationship is not always the rule in the automobile industry. In the same reference, there is another quote from a supplier to Toyota, saying '*Toyota helped*

[2]Liker, J. K. and T. Y. Choi (2004). "Building deep supplier relationships." *Harvard Business Review, 82*, 12, 104–113 (2004).

Table 1 Notation

• x_i: Firm i's cumulative amount of resources invested in new product development, $i = 1, 2$ (state variable); the manufacturer, $i = 1$
• u_i: Firm i's resources invested in new product development at t, $i = 1, 2$ (control variable); the manufacturer, $i = 1$
• $\bar{u}_i$: Limit on firm i's resources that can be invested at t
• a_i: Parameter that determines the firm i's profitability
• ρ_i: Parameter that determines the cost to use the firm i's resources

us dramatically improve our production system. We started by making one component, and as we improved, [Toyota] rewarded us with orders for more components. Toyota is our best customer.'

1.2.2 The Single Player-Dominating Decision-Making Structure: A Coercive Case

For our first case, we develop an optimal control theory model that describes the manufacturer-dominating supply chain. In general, the manufacturer has a larger bargaining power vis-à-vis its supplier, implying that the manufacturer can exert much larger decision-making power than its supplier. We embody the manufacturer's domination in the objective function of the optimal control model in *P1_M*. Table 1 explains the variables and parameters in the model.

P1_M:

$$Maximize \int_0^T \{(x_1 + x_2)(a_1 - (x_1 + x_2)) - \rho_1 u_1^2\} dt \tag{1}$$

$$Subject\ to\ \dot{x}_1 = u_1 \tag{2}$$

$$u_1 \leq \bar{u}_1 \tag{3}$$

$$\dot{x}_2 = u_2 \tag{4}$$

$$u_2 \leq \bar{u}_2 \tag{5}$$

$$x_1(0) = \alpha_1, x_2(0) = \alpha_2 \tag{6}$$

P1_M is the manufacturer's decision problem: This optimal control model consists of an objective function and seven constraints. We explain each of the components in detail.

- $\int_0^T \{(x_1 + x_2)(a_1 - (x_1 + x_2)) - \rho_1 u_1^2\} dt$: In the objective function (1), the profit maximization is done from the manufacturer's perspective and only the cost of

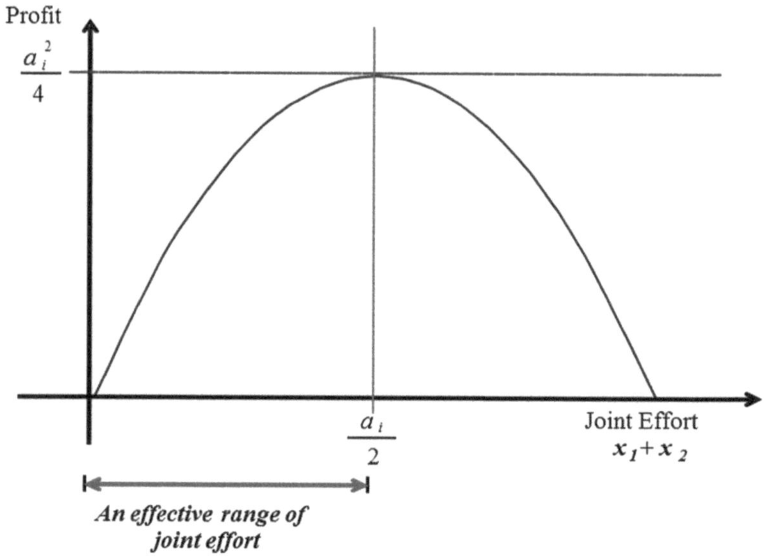

Fig. 2 Profit function

using the manufacturer's resource is included in the model. That is, when the manufacturer can wield its dominating bargaining power, it can make a decision on how the supplier's resource should be utilized, but does not take into account the cost of using it. Although this kind of decision-making structure or arrangement is unfair from the supplier's viewpoint, in reality it can happen when the supply chain partners have extremely skewed bargaining powers.

(a) $(x_1 + x_2)(a_1 - (x_1 + x_2))$: The first part in the integrand of the objective function is the manufacturer's profit, jointly determined by the manufacturer's effort and the supplier's at the same time, and concave in the joint effort, $x_1 + x_2$: This format of profit function is widely used, in particular when competing or coordinating players have to devote effort jointly in order to create common ground (e.g., a public good) on which each of them can earn its own profit (see Fig. 2). That is, the manufacturer needs the supplier's effort, i.e., investment in new product development, in order to generate its own profit, and vice versa. In this sense, it is important for the supply chain partners to coordinate with each other. As we can see in Fig. 2, there might be a technical issue: There is an effective range of joint effort, during which a supply chain partner's profit increases as the joint effort level increases. Once the joint effort level exceeds the range, the profit decreases even when the joint effort increases. Although this kind of phenomenon can happen if we allow 'diseconomies of scale' in terms of the joint effort, in order to avoid any unnecessary (and unrealistic) complications in the ensuing discussion, we would like to consider only the cases, where the joint effort is within the effective range.

(b) $\rho_1 u_1^2$: Such a joint effort requires cost, i.e., each supply chain partner needs to pay its fair share of cost. However, as alluded to in the automobile industry, it does happen that the manufacturer makes decisions on how to utilize the supplier's resources when it has dominating bargaining power vis-à-vis the supplier. When this kind of situation indeed happens, from the manufacturer's perspective, the supplier's resources are free and the cost to utilize them is not included in the manufacturer's decision problem. That is why there is only one cost element, $\rho_1 u_1^2$, in *P1_M*.

- $\dot{x}_i = u_i$: In *P1_M*, (2) and (4) show the dynamics of the cumulative resource investments. Without loss of generalizability, we assume that the effort level is scaled so that investing one unit of resource increases one unit of effort.
- $u_1 \leq \bar{u}_1$: (3) and (5) specify the resource availability that can be utilized at t for each of the players. For instance, the manufacturer cannot spend more than $\bar{u}_1$ at t on developing a new product.
- $x_i(0) = \alpha_i$: (6) presents the initial values of the state variables.

With *P1_M*, we formulate the Hamiltonian as in (7) and solve the optimal control model as in Appendix "A Single Player-Dominated (Coercive) Case". The Hamiltonian or Lagrangian is structured as follows:

$$\begin{aligned} L &= H \\ &= (x_1 + x_2)(a_1 - (x_1 + x_2)) - \rho_1 u_1^2 + \lambda_1 u_1 + \lambda_2 u_2 + w_1(\bar{u}_1 - u_1) + w_2(\bar{u}_2 - u_2) \end{aligned} \tag{7}$$

Now, the profit of the weaker player in the supply chain, i.e., the supplier, is completely determined by the solutions to *P1_M*: '∗' means the value is part of the optimal solution determined in the manufacturer's problem. In the single player-dominating case, where the manufacturer wields excessively large bargaining power, the supplier can only react passively, allowing the manufacturer to decide how to utilize the supplier's resources. Therefore, the supplier's decision problem *P1_S* is simply to calculate its profit by plugging in its profit function the solved decision variables, both state and control variables:

P1_S:

$$\text{Supplier's profit} = \int_0^T \left\{ \left(x_1^* + x_2^*\right)\left(a_2 - \left(x_1^* + x_2^*\right)\right) - \rho_2 \left(u_2^*\right)^2 \right\} dt \tag{8}$$

1.2.3 The Balanced Decision-Making Structure: A Balanced Case

In this section, we develop an optimal control theory model for the second case, the balanced decision-making structure. For this case, a shared decision-making or a balanced allocation of bargaining power between supply chain partners is assumed.

We already discussed cases about a carmaker's relationship and coordination in developing new models with its suppliers. Even if the manufacturer can wield a dominating bargaining power, it might not want to make a decision unilaterally since it knows that such a biased decision-making will not improve the profitability not only for the supplier, but also for the manufacturer itself.

The most significant difference from the single player-dominated case is the objective function. That is, in the objective function of the balanced decision-making structure, the profit to be maximized is the sum of those for the supplier and the manufacturer. In addition, the cost to utilize the supplier's resources is also included in the objective function, implying that the supplier's resources are not free any longer even to the manufacturer. This arrangement in the objective function implies that the manufacturer and the supplier are now doing joint decision-making, i.e., the balanced decision-making structure. That is, we embody the balanced decision-making scheme by taking into account the costs and benefits of the supplier fully in the objective function. *P*2 is the optimal control theory model for this case.

*P*2:

$$\textit{Maximize} \int_0^T \{(x_1+x_2)(a_1+a_2-2(x_1+x_2))-\rho_1 u_1^2-\rho_2 u_2^2\}\mathrm{d}t \tag{9}$$

$$\textit{Subject to } \dot{x}_1 = u_1 \tag{10}$$

$$u_1 \le \bar{u}_1 \tag{11}$$

$$\dot{x}_2 = u_2 \tag{12}$$

$$u_2 \le \bar{u}_2 \tag{13}$$

$$x_1(0)=\alpha_1, x_2(0)=\alpha_2 \tag{14}$$

The relevant Hamiltonian or Lagrangian is as follows:

$$\begin{aligned} L &= H \\ &= (x_1+x_2)(a_1+a_2-2(x_1+x_2))-\rho_1 u_1^2 \\ &\quad -\rho_2 u_2^2+\lambda_1 u_1+\lambda_2 u_2+w_1(\bar{u}_1-u_1)+w_2(\bar{u}_2-u_2) \end{aligned} \tag{15}$$

In *P*2, (10)–(14) are exactly the same as (2)–(6). Using the Hamiltonian (15), we derive the analytical solution to *P*2, the procedures of which are well documented in Appendix "A Balanced Decision-Making Case".

1.3 Analysis of the Model

In the previous section, we've developed optimal control theory models for the two cases, the coercive (single player-dominating) and the balanced decision-making situation. Since the detailed mathematical derivations are completely listed in Appendix "A Single Player-Dominated (Coercive) Case" and "A Balanced Decision-Making Case", we do not have to repeat the problem solving procedures here. Rather, in this section, we would like to discuss the dynamics of the solutions by depicting them in figures: We focus on how to analyze the patterns of resource allocation dynamics and where the benefits of the balanced decision-making come from. For the single player-dominating case, we show why the manufacturer (with the superior bargaining power) always uses up the supplier's resources first before it starts using its own: We prove the supply chain participants' behavioral dynamics by analyzing their marginal valuations reflected in costate variables and Lagrangian multipliers. We conduct a similar analysis for the balanced case as well.

1.3.1 Dynamics of the Single Player-Dominated Case: The Coercive Case

In Appendix "A Single Player-Dominated (Coercive) Case", we have detailed the solution procedures for the single player-dominating case: We assume the dominant player is the manufacturer, reflecting the reality in general. We have paid special attention to the behaviors of λ_i and w_i in order to understand under what circumstances the sole decision-maker (the manufacturer in our case) uses the resources to their limits, i.e., the dynamics of $u_i^* \leq \bar{u}_i$. Our analysis indicates that since the supplier's resources are essentially free to the manufacturer, assuming the resources are valuable, the manufacturer always consumes the supplier's resources to the limit, i.e., $u_2^* = \bar{u}_2$. In addition, the result shows that the relative magnitude between $\lambda_1(0)$ and $2\rho\bar{u}_1$ is critical to determining whether the manufacturer uses up its own resources completely or not: $\lambda_1(0)$ is the shadow price or marginal value of the unit resource at the beginning of the decision time horizon; since ρ is a cost coefficient and $\bar{u}_1$ is the limit of the manufacturer's resources, $2\rho\bar{u}_1$ somehow represents the total cost to utilize its own entire resources.

There are two cases:

- Case 1: $\lambda_1(0) < 2\rho\bar{u}_1$, i.e., the marginal value of unit resource is less than a measure related with the total cost to use up the manufacturer's entire resources.
- Case 2: $\lambda_1(0) > 2\rho\bar{u}_1$, i.e., the marginal value of unit resource is larger than a measure related with the total cost to use up the manufacturer's entire resources.

Figure 3 describes Case 1. As in Appendix "A Single Player-Dominated (Coercive) Case" and Fig. 3, we see that under Case 1, the manufacturer never consumes its own resources completely, while it always uses up the supplier's: $u_1^* < \bar{u}_1$ and $u_2^* = \bar{u}_2$ throughout $0 \leq t \leq T$. It is, therefore, reasonable to infer that under this circumstance, it might be unavoidable for the resources to be utilized

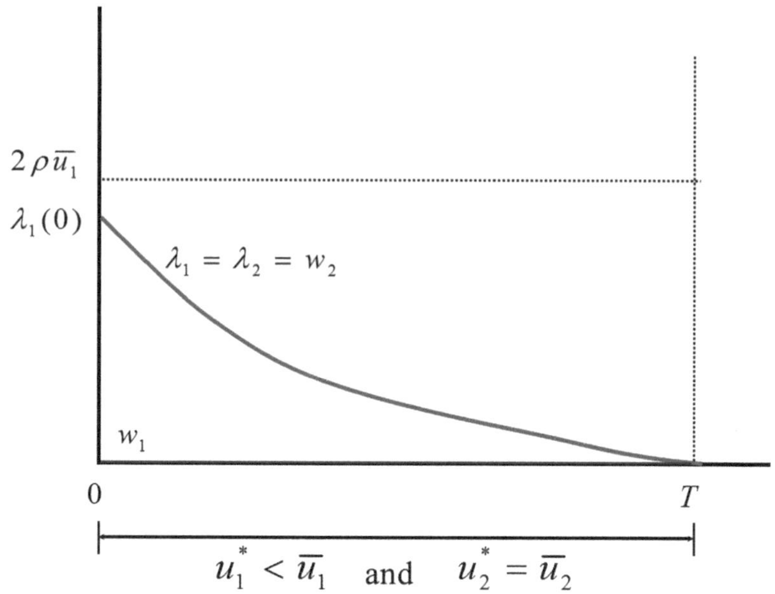

Fig. 3 Dynamics of resource utilization when $\lambda_1(0) < 2\rho\bar{u}_1$

inefficiently and ineffectively. Such inefficient resource utilization easily causes suboptimality of performance in the supply chain as a whole.

In Case 2, there can exist $\hat{t}$ such that $u_1^* = \bar{u}_1$ for $0 \le t < \hat{t}$ and $u_1^* < \bar{u}_1$ for $\hat{t} \le t < T$: see Fig. 4. Appendix "A Single Player-Dominated (Coercive) Case" proves that

$$\hat{t} = \sqrt{\frac{\lambda_1(0) - 2\rho\bar{u}_1}{\bar{u}_1 + \bar{u}_2}}. \tag{16}$$

(16) indicates the following:

(a) $\dfrac{d\left(\frac{\lambda_1(0)-2\rho\bar{u}_1}{\bar{u}_1+\bar{u}_2}\right)}{d\bar{u}_1} = -\dfrac{\lambda_1(0)+2\rho\bar{u}_2}{(\bar{u}_1+\bar{u}_2)^2} < 0$, implying that the larger the manufacturer's resource limit, the shorter the $\hat{t}$, i.e., the shorter the period during which $u_1^* = \bar{u}_1$, other things being equal. That is, when the manufacturer's resources are abundant, it becomes less likely that the manufacturer consumes its resources fully.

(b) $\dfrac{d\left(\frac{\lambda_1(0)-2\rho\bar{u}_1}{\bar{u}_1+\bar{u}_2}\right)}{d\bar{u}_2} = \dfrac{-\lambda_1(0)+2\rho\bar{u}_1}{(\bar{u}_1+\bar{u}_2)^2} < 0$, implying that the larger the supplier's resource limit, the shorter the $\hat{t}$, i.e., the shorter the period during which $u_1^* = \bar{u}_1$, other things being equal. That is, when the supplier's resources are abundant, it becomes less likely that the manufacturer consumes its resources fully because it would be sufficient to use up the supplier's resources first.

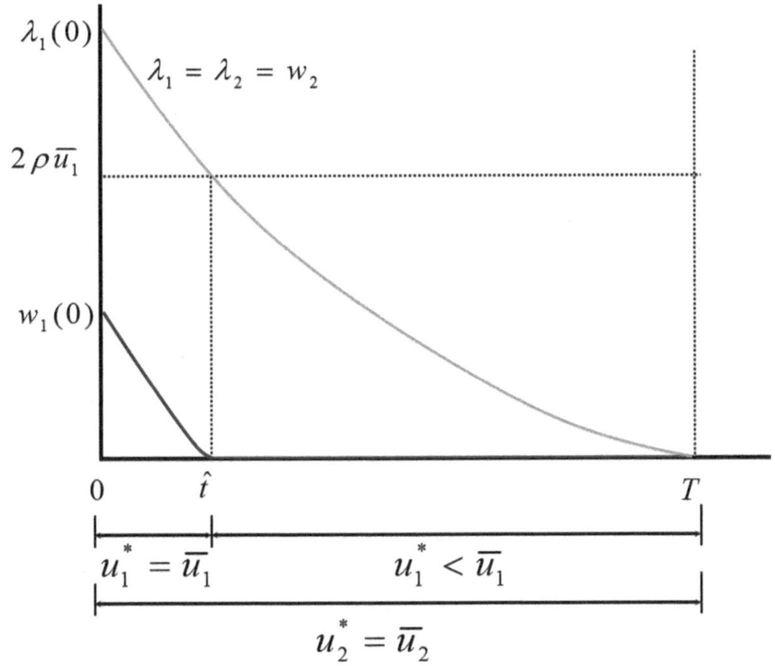

Fig. 4 Dynamics of resource allocation when $\lambda_1(0) > 2\rho\bar{u}_1$

Which situation is more likely to occur in reality, $\lambda_1(0) < 2\rho\bar{u}_1$ or $\lambda_1(0) > 2\rho\bar{u}_1$? It might depend on the specific context in point. But, $\lambda_1(0)$ is the shadow price or marginal value of the unit resource, while $2\rho\bar{u}_1$ somehow represents the total cost to utilize the full resources, and therefore, in general, $\lambda_1(0) < 2\rho\bar{u}_1$ seems more likely to occur. This observation implies that it is more likely to have $u_1^* < \bar{u}_1$ and $u_2^* = \bar{u}_2$ throughout $0 \leq t \leq \mathrm{T}$, i.e., the manufacturer always consumes the supplier's resources completely, whereas it never uses up its own resources, regardless of the relative efficiency or effectiveness between the two different types of resources. Hence, it is more likely to waste valuable resources.

1.3.2 Dynamics of the Balanced Decision-Making Case: The Balanced Case

Now, we would like to analyze the dynamics of resource allocation for the balanced decision-making case. Again, the details of mathematical derivations are well documented in Appendix "A Balanced Decision-Making Case". Similar with our analysis for the single player-dominating case, the precise dynamics of resource allocation for the balanced case are dependent on the relative magnitudes among $\rho_1\bar{u}_1$, $\rho_2\bar{u}_2$, and $\lambda(0)$ (see Fig. 16 and Table 8).

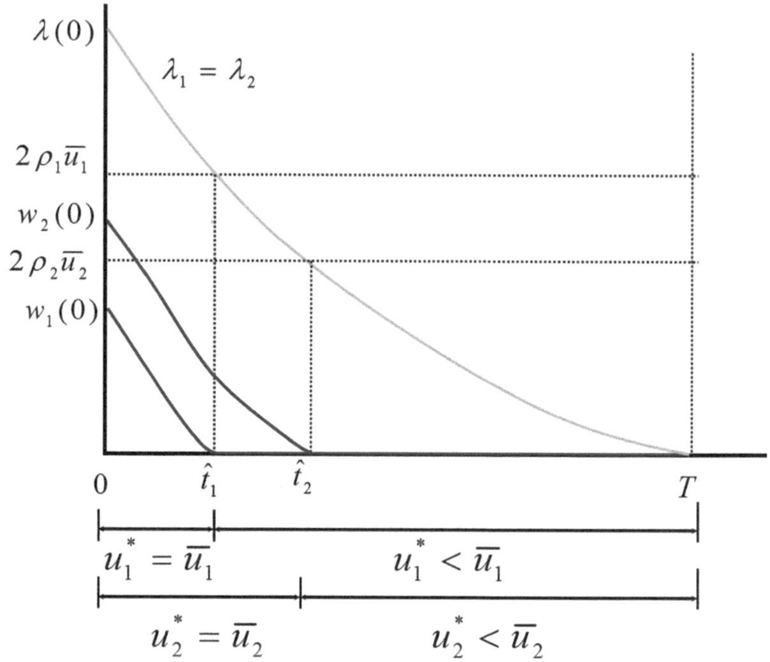

Fig. 5 Dynamics of the resource utilization for the balanced case: $\lambda(0) > 2\rho_1\bar{u}_1 > 2\rho_2\bar{u}_2$

Figure 5 depicts the case where $2\rho_2\bar{u}_2 < 2\rho_1\bar{u}_1 < \lambda(0)$. Under this condition, $\hat{t}_1$ and $\hat{t}_2$ can be determined as in (72) and (77):

$$\hat{t}_1 = \sqrt{\frac{\lambda_1(0) - 2\rho_1\bar{u}_1}{2(\bar{u}_1 + \bar{u}_2)}}, \tag{17}$$

$$\hat{t}_2 = T - \frac{\rho_2\bar{u}_2}{\rho_1\bar{u}_1}(T - \hat{t}_1). \tag{18}$$

We can make a similar interpretation for $\hat{t}_1$ with that for $\hat{t}$ in (16). It is noteworthy that $\hat{t}_2$ is determined mainly by two elements: the difference between $\hat{t}_1$ and T as expressed in $(T - \hat{t}_1)$, and the relative efficiency between the two different resource utilizations as represented in $\frac{\rho_2\bar{u}_2}{\rho_1\bar{u}_1}$.

The biggest difference between the single player-dominating and the balanced case is that the decision-makers in the balanced case use resources by taking into account the relative efficiency of the resources, whereas in the coercive case, the resources owned by the player with less bargaining power are used up completely regardless of their efficiency vis-à-vis that of the resources owned by the dominant decision-maker.

1.4 Numerical Examples

To cast our analysis in the real-world context, we conduct a numerical analysis. Again let us consider the automobile industry. The setting can be described as follows. The carmaker gains benefit as it becomes more capable of developing a new model faster, i.e., reducing the new product development (NPD) lead time. In order to improve its NPD speed, the carmaker needs support from its supplier. Now, we have the following one-to-one relationship.

- x_1: the cumulative investment of resources by the carmaker in enhancing its NPD capability,
- x_2: the cumulative investment of resources by the supplier in enhancing the carmaker's NPD capability,
- $x_1 + x_2$: the total cumulative investment by both the carmaker and its supplier, which reflects or represents the supply chain's overall NPD capability, and
- $(x_1 + x_2)(a_i - (x_1 + x_2))$: the profit the carmaker $(i = 1)$ or the supplier $(i = 2)$ earns when the NPD capability is $x_1 + x_2$ at t.

Table 2 presents the parameter values used in the numerical analysis: All of them have a unit in million dollars, $100 MM, except for T, which is in years.

Using the parameter values in Table 2, we conduct numerical analyses for the two different scenarios. Numerical analysis results are reported in Table 3, which we recapitulate as follows:

- Total payoff (combining the manufacturer's and the supplier's together) of the balanced case is much larger than that of the coercive case, i.e., the single player-dominating case.
- It is intriguing to note that the balanced case total payoff is larger, despite that the total cumulative investment of the balanced case is smaller than that of the

Table 2 Parameter values

$a_i(i = 1, 2)$	ρ_1	ρ_2	$\bar{u}_1$	$\bar{u}_2$	α_1	α_2	T
1	120	100	0.03	0.03	0	0	10

Table 3 Numerical analysis results (in $100 MM)

Single player-dominated (coercive)		Balanced decision-making	
Total payoff	20.90	Total payoff	24.44
Manufacturer's payoff	14.19	Manufacturer's payoff	13.33
Supplier's payoff	6.71	Supplier's payoff	11.11
Cumulative investment: total	0.39	Cumulative investment: total	0.32
Cumulative investment: manufacturer	0.09	Cumulative investment: manufacturer	0.15
Cumulative investment: supplier	0.30	Cumulative investment: supplier	0.17

coercive case. It implies that the supply chain partners utilize resources much more effectively under the balanced decision-making structure than under the single player-dominating structure. We infer that when it comes to utilizing its own resources, each partner in the supply chain knows better than the other. Under the balanced case, the manufacturer must increase its investment, while the supplier can reduce its spending significantly.

- When we look at the individual profits, we notice that the supplier's profit increases significantly as shifting from the coercive to the balanced case, whereas the manufacturer's actually decreases a little bit. Here, we can see a potential dilemma faced by the manufacturer. When the manufacturer is exerting its huge bargaining power without any restraints, it is able to enjoy higher profit, although the supply chain as a whole is operating at an inefficient level. In order to make the entire supply chain more efficient, the manufacturer need change its decision-making structure so that a more balanced decision-making becomes possible. By doing so, the entire supply chain performs better, but at the cost of the manufacturer's profit. Is there any solution that can eliminate or at least mitigate such a dilemma faced by the manufacturer? One possible way is for the supply chain partners, the manufacturer and its supplier, to agree upon a scheme to share the total profit in a more equitable manner; e.g., each player's payoff under the balanced case should be strictly larger than that under the coercive case. If they can find a mechanism that measures how much contribution is made by each player to enlarging the total profit, the profit sharing issue under the balanced case can be much simpler. Table 4 summarizes the incremental changes as the supply chain shifts from a single player-dominating to a balanced decision-making structure.

Figure 6 depicts the dynamics of total payoffs, whereas Fig. 7 shows the dynamics of the manufacturer's payoff and Fig. 8 the supplier's. For the manufacturer, it is always better to have a coercive decision-making structure (of course, since the manufacturer is the dominant player). On the contrary, for the supplier, it is always better to have a balanced decision-making structure, where both the manufacturer and the supplier share the decision-making power equitably. One interesting observation is on the dynamics of total payoffs in Fig. 6. Although the total payoff under the balanced decision-making structure becomes much better than that under the single player-dominating structure eventually, it actually stays

Table 4 Comparison between the two cases

[(Balanced case value—Coercive case value)/Coercive case value]%	
Total payoff	17%
Manufacturer's payoff	−6%
Supplier's payoff	66%
Cumulative investment: total	−18%
Cumulative investment: manufacturer	67%
Cumulative investment: supplier	−43%

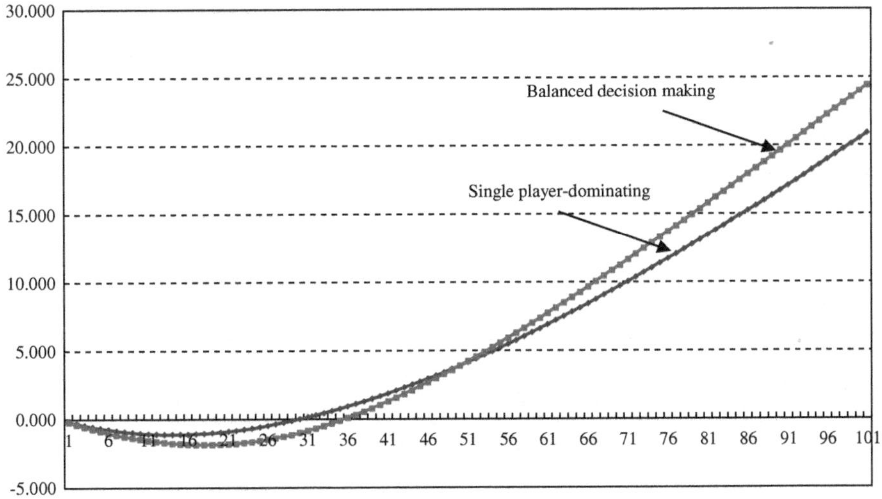

Fig. 6 Total payoff

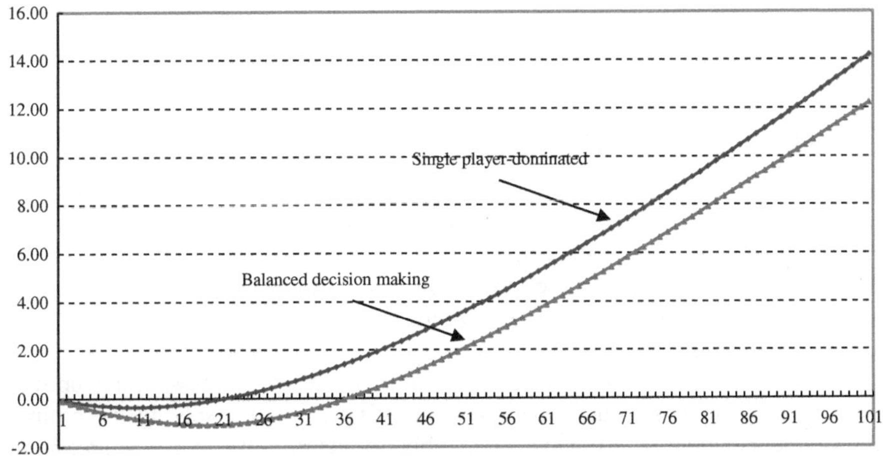

Fig. 7 Payoff of the dominating player, i.e., the manufacturer

lower than the coercive payoff almost until the half of the decision time horizon: That is, the total payoff under the coercive case is larger than that under the balanced case in the early period of the decision time horizon. It is due to that the balanced decision-making structure requires relatively more initial investment from the supply chain partners, which can be utilized fully in the later period: This kind of early commitment (of investment) is possible because the supply chain partners have trust in each other. As such, we can infer that if the supply chain partners have

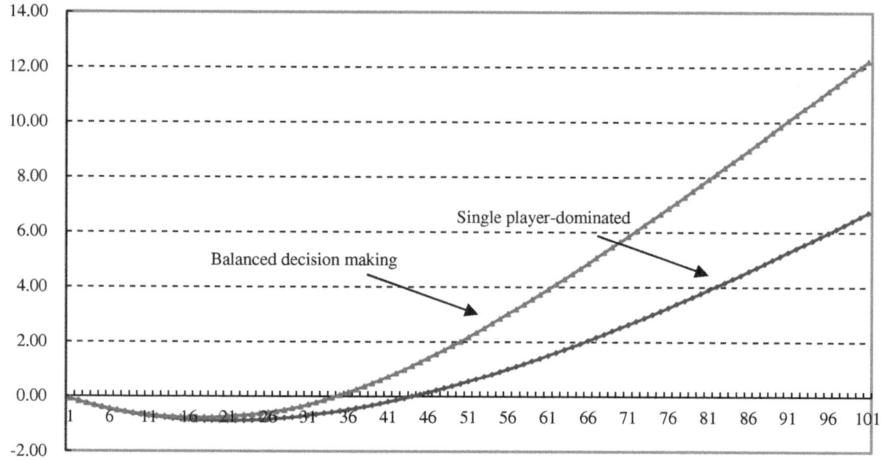

Fig. 8 Payoff of the weaker player, i.e., the supplier

a relatively short decision time horizon (i.e., a myopic perspective), the balanced decision-making structure might not be the ideal choice, since it may generate less total profit than the coercive case does.

1.5 Managerial Implications

Our primary objective in this section is to understand the detailed dynamics of resource allocation and profit generation under the two important supply chain decision-making mechanisms, i.e., the single player-dominating (coercive) and the balanced decision-making structure. Essentially, we would like to answer a question, '*Where comes the benefit of the balanced decision-making between supply chain partners and how can we analyze its dynamics?*' by developing optimal control theory models to describe two contrasting cases. By solving the optimal control theory models, we have been able to identify the exact sources and dynamics of inefficiency or efficiency of each of the decision-making structures.

Under the arrangement of a single player-dominating decision-making, the dominant partner always consumes the less-powerful partner's resources completely before it starts using its own resources. It is due to that the less-powerful partner's resources are essentially free to the dominant player in the supply chain: When something is free, a human being tends to use it fully without paying due attention to whether such usage is efficient or economical from the entire system's perspective. That is, under the single player-dominating case, it is unavoidable to waste resources to a great extent, and such resource waste causes the entire supply chain to perform suboptimally.

Under the balanced decision-making structure, it costs to use not only the dominant partner's resources, but also the less-powerful player's resources: In the objective function, both the manufacturer's and the supplier's benefits and costs are equally represented. Our analysis indicates that under this balanced decision-making structure, resource utilization is disciplined in that the decision-makers determine how much of each player's resources should be utilized by taking into account their relative efficiency. Such disciplined resource utilization enables the entire supply chain to perform more optimally.

Numerical examples clearly show the detailed dynamics of the resource allocation and profit generation for the two cases. Some of the key results are as follows: Total payoff of the balanced case is much larger than that of the single player-dominating case, despite that the total cumulative investment of the balanced case is smaller, indicating that the supply chain partners utilize resources more effectively under the balanced decision-making structure, but the supplier's profit increases significantly as shifting from the coercive to the balanced case, whereas the manufacturer's actually decreases a little bit; to fix this dilemma, the manufacturer and its supplier must agree upon a scheme to share the total profit in a more equitable manner; if they can find a mechanism that measures how much contribution is made by each player to enlarging the total profit, the profit sharing issue under the balanced case can be much simpler.

2 Supply Chain Coordination[3]

Firms' long-term competitiveness depends on their ability to carry out product and process innovation, which enables them to appropriate extraordinary rents for a certain period of time, say until another innovation makes it obsolete. Studies on supply chain management have emphasized the importance of a long-term strategic relationship between a manufacturing firm and its suppliers. The fundamental assumption underlying this emphasis is that the long-term relationship makes both the manufacturer and the suppliers better off than when there is no such relationship. Therefore, we can infer that in order to be sustainable, the supplier–manufacturer relationship must result in enhancing the profitability of the supplier as well as the manufacturer itself. The essence of this relationship is concerned with coordination between the two participants.

As alluded already, a sustaining supplier–manufacturer relationship cannot be possible without tangible benefits accrued to both partners. Such phenomenon was described as Pareto-improving by Iyer and Bergen (1997). They postulated that with the arrangement of QR (quick response) between a buyer and a supplier, the buyer can always get benefits, but it might not be attractive to the supplier. Based

[3]This section is a refined version of Kim, B. (2000) "Coordinating an innovation in supply chain management." *European Journal of Operational Research, 123* (3), 568–584.

on the empirical studies, they further suggested that the buyer's service-level commitments could reduce the burden as well as uncertainty the supplier should bear. In addition, other industry practices to overcome the unbalanced benefit were presented such as contractual commitment between the manufacturer and the supplier, and cooperative advertisements.

Coordination is often defined as the process of managing dependencies among activities. There are different kinds of dependencies associated with aspects such as shared resources, task assignments, producer and consumer relationships, and design for manufacturability. We consider manufacturer's supporting supplier innovation as a way to deal with interorganizational coordination, e.g., managing dependence between manufacturing firm and its supplier(s): Supplier innovation could relate to improvements in quality, yield, delivery time, and also supply cost (Hartley et al. 1997; Bernstein and Kök 2009). That is, in order to make it attractive for the supplier to participate in the supply chain coordination, the manufacturer must convince the supplier that such relationship will benefit the supplier, too. The manufacturer's support can be regarded as an incentive for the supplier.

If we view the manufacturing firm and its supplier as consisting in a network, their network profit can be defined as the profit achieved through the coordination between themselves, but before allocated to each of them individually (Jeuland and Shugan 1983). Thus, we can suggest the first condition for a sustainable supplier–manufacturer coordination: That is, the network profit attained through the coordination must be larger than the sum of uncoordinated individual profits. Since the network profit is before the allocation to each individual participant, an optimal strategy at the network level might not necessarily be optimal to each individual firm in the supply chain. Thus, the second condition for a sustainable supplier–manufacturer coordination is that at least in the long run, each individual participant in the supply chain must perceive its allocated profit is larger than that it could attain with no such coordination.

In this section, the coordination is reflected in the manufacturing firm's support for the supplier's innovation activities: We specifically focus on a particular type of supply chain coordination, i.e., supplier–manufacturer coordination for supplier innovation. The fundamental premise is that the manufacturer's subsidy for its supplier's innovation can eventually lead to supply cost reduction and thus the manufacturer's own product price. The supplier's innovation is assumed to affect the supply quality and delivery time, both of which are combined to determine the supply cost. We will show that although the manufacturer's subsidy for its supplier can be an incentive for the supplier's innovation, a simple subsidy might not be enough to induce the supplier's cooperation under a certain market demand structure.

2.1 Model Formulation

In order to specifically focus on the intricate relationship of manufacturing firm's supporting the supplier innovation, we model a supply chain consisting of only one manufacturer and its supplier (Dai et al. 2012). Analytical and managerial implications from this sparse model can be extended to more complicated situations.

Figure 9 depicts the present model situation. After receiving supplies, e.g., raw materials or intermediate inputs, from the supplier, the manufacturing firm produces its final products and sells them to the market. As the arrow from the manufacturer to the supplier indicates, the primary decision-making is concerned with whether the manufacturer should support its supplier's innovation that can lead to the reduction in the supply cost.

As shown in Fig. 9, there is another important element that can affect the supplier–manufacturer relationship, which is the Market Demand. In the ensuing analysis, we consider two different cases in which the market demand structure behaves. In the first case, we assume the market demand for the final product is constant. It is equivalent to a premise that the manufacturing firm is in a competitive or perfect market with a fixed production capacity: This situation is comparable with the competitive market condition, where a single firm cannot influence the prevailing market price, and therefore, at least for the short run, the market demand for that particular firm is constrained by the production capacity, assuming that the marginal revenue is larger than the marginal production cost. From the manufacturer's viewpoint, there is little incentive to cut its price since the price cut will surely be followed by other firms' given that the market is competitive. Therefore, under such a circumstance, the market demand for the firm's product will stabilize around a fixed quantity, e.g., the firm's production capacity. On the other hand, if the manufacturing firm can reduce the cost for the final product, it can increase its profit to the extent of cost saving.

We relax this assumption in the second case, where the market demand for the manufacturer's product is a function of the product price set by the manufacturing firm. That is, the lower the final product's market price, the larger the market demand for the product. This situation might be comparable with an oligopolistic or

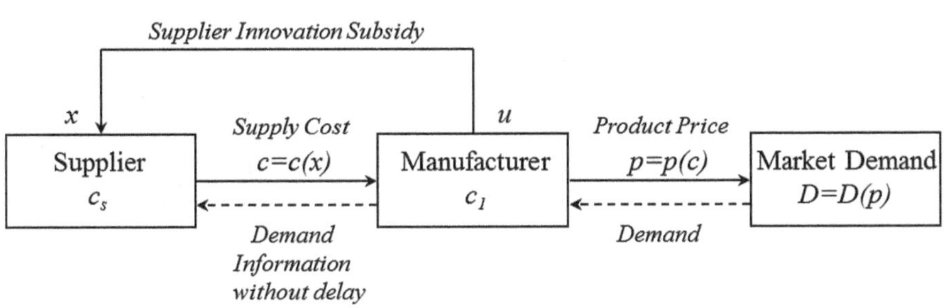

Fig. 9 A simple model structure

Table 5 Comparison of the two models

	Simple model	Extended model
Manufacturing capacity	• Fixed capacity, D (much smaller than the aggregate market demand)	• Flexible capacity (insignificant fixed cost, e.g., changeover cost)
Demand structure	• Competitive market • Price, p, given • *Deterministic demand*	• Monopolistic or oligopolistic market (imperfect market) • D as a function of p set by the manufacturer • *Deterministic demand*
Decision impact	• Constant revenue due to price given • Cost reduced • Profit increase through cost reduction only • *Supplier subsidy as profit loss to the manufacturer*	• Revenue change via price change • Cost reduced • Profit increase through cost reduction and revenue increase • *Supplier subsidy as profit loss to the manufacturer*
Demand information	*Insignificant delay between supplier and manufacturer*	
Analysis	*Comparing profits for with and without coordination (subsidy)*	

monopolistic market condition, where a firm can increase the market demand by lowering the product price. How sensitive is the market demand response to the price change will be an important component to be determined in the analysis. Table 5 compares the two models analyzed in this section.

In addition, for analytical simplicity without loss of generality, we assume that there is little information delay between the manufacturer and its supplier. Thus, we do not consider any inefficiency due to an information distortion in the supply chain.

2.1.1 A Simple Model: The Case of Constant Market Demand

The first simple model assumes constant production capacity and a given market price for the product. In this situation, the manufacturing firm cannot unilaterally lower its price to attract more demand. But, it can try to reduce the cost so as to increase the profit. The production capacity of the manufacturing firm, i.e., essentially the market demand for the firm, is given as D at time $t \in [0, T]$, and the market price for the product is given as p. Manufacturer's total unit cost of the product consists of two parts, internal production cost (c_1) and supply cost (c_2) charged by the supplier. Therefore, the unit profit is $p - c_1 - c_2$ and the total profit realized at t is $D(p - c_1 - c_2)$. Since we assume that p and D are exogenously determined, the only way for the manufacturer to earn more profit is to reduce either c_1 or c_2. In this section, we are interested in the collaborative relationship between a manufacturing firm and its supplier, and thus focus on supplier innovation to reduce c_2.

From the supplier's point of view, there would be little incentive to lower c_2 unless such innovation enables the supplier to earn more profit. If the manufacturing firm wants to increase the unit profit and the only remaining alternative is to reduce c_2, then it might offer a certain amount of support or subsidy for the supplier to innovate.

In the ensuing analysis, it is assumed that the supplier takes on projects or experiments as innovation activities to reduce the supply cost. Thus, the manufacturer's help for the supplier's innovation is to support or subsidize the innovation projects. Let us denote $u(t)$ to represent the number of projects the manufacturing firm subsidizes for the supplier's innovation at t. We further impose a constraint $0 \leq u(t) \leq a$ which implies that the manufacturing firm cannot support more than a projects at any given time: The exact size of a depends on the firm's internal resource availability as well as its capability. Since an accurate determination of a is not critical to our discussion in this section, a presumption is made that a is exogenously assessed so as to be feasible to the problem.

For analytical simplicity without loss of generality, $u(t)$ is considered as a continuous variable. In addition, it is also presumed that the manufacturer's total cost to support one such project is α.

Let us denote $x(t)$ as the cumulative number of projects supported by the manufacturer up to t, i.e., $x(t) = \int_0^t u(t)\mathrm{d}t$ and $\frac{\mathrm{d}x(t)}{\mathrm{d}t} = \dot{x}(t) = u(t)$. Now, we can express c_2 as a function of $x(t)$, satisfying $\frac{\mathrm{d}c_2}{\mathrm{d}x} < 0$. For analytical tractability, a specific functional form is adopted, based on the well-established learning curve model (Yelle 1979), so that $c_2 = cx^m$, where c is the base supply cost, $m = \frac{ln\phi_m}{ln\chi}$, $\phi_m = 1 - \Xi$, and $0 \leq \Xi \leq 1$ the learning rate, which is determined by the supplier's innovation capability. For instance, if $\Xi = 0.05$ and $\chi = 2$, c_2 is reduced by 5% each time $x(t)$ doubles while $x(t) \geq 1$. Although it can be conceptually easier to follow, adopting the learning curve formula for c_2 is not essential for the ensuing analysis: c_2 can be further simplified by imposing a more succinct constraint $m < 0$, not involving the learning rate. Table 6 shows the summary of variables and parameters for the first simple model.

Employing the variables and parameters defined so far, we can formulate the first simple model as follows:

*P*3:

$$Maximize\ J = \int_0^T [D(p - c_1 - c_2) - \alpha u]\mathrm{d}t \tag{19}$$

$$Subject\ to\ \dot{x} = u \tag{20}$$

$$0 \leq u \leq a, x(0) = x_0 = 1 \text{ specified.} \tag{21}$$

Table 6 Summary of variables and parameters for the simple model

D: production capacity (market demand) at t
p: market price of the finished good
c_1: internal unit production cost
$c_2 = cx^m$: supplier's unit supply cost to the manufacturer
c: supply unit cost base
$m = \frac{ln\phi_m}{ln\chi}$
ϕ_m: capturing supplier's innovation potential
χ: capturing 'how fast the cost reduction occurs'
$x(t)$: cumulative number of innovation projects supported by the manufacturer
$x(0) = x_0 = 1$ assumed
$u(t)$: number of projects the manufacturer subsidizes at t
a: maximum number of projects the manufacturer can afford at any given time
α: manufacturer's cost associated with an innovation project
$[0, T]$: decision horizon

As mentioned already, α is the manufacturer's total cost associated with supporting one innovation project. We further assume that out of this α, γ_m is the net amount actually paid to the supplier: $0 \leq \gamma_m < \alpha$ is imposed. Thus, $\alpha - \gamma_m$ is the amount that is paid to the external entities outside the supply chain.

$x(0) = 1$ implies that at the beginning, the supplier has already conducted one project (or experiment) which enables it to provide the manufacturer with supplies at a unit cost of c. It is premised that as the number of projects or experiments the supplier performs increases, the supply cost is reduced: The speed of cost reduction depends on the parameters constituting the cost function. For the purpose of the analysis, a time value of the profit is assumed insignificant, and thus, the discounting factor is not considered.

To solve $P3$, we make use of the maximum principle in optimal control theory, the associated Hamiltonian being

$$H = D(p - c_1 - cx^m) - \alpha u + \lambda u. \tag{22}$$

In order to obtain an optimal solution, we take a partial differentiation of (22) with respect to u. Because H is a linear function of u, we cannot use the usual optimization criterion, i.e., $\frac{\partial H}{\partial u} = 0$, nor is it necessary to check the sufficient condition for optimality. Rather, we first obtain $\frac{\partial H}{\partial u} = -\alpha + \lambda$ and find out a binary solution,

$$u^* = \begin{cases} a, \text{ if } \lambda \geq \alpha \\ 0, \text{ if } \lambda < \alpha \end{cases} \tag{23}$$

see Appendix 3 for the discontinuous control variable.

Applying the necessary conditions of the maximum principle,

$$\dot{\lambda} = -\frac{\partial H}{\partial x} = cm\,Dx^{m-1}. \tag{24}$$

Since $m<0$, $\dot{\lambda}<0$ always holds. We know additional characteristics of λ: (i) It represents the marginal value of $x(t)$, (ii) $\lambda(T) = 0$ because $x(T)$ is left free, and (iii) it is a continuous function. Therefore, it must be that $\lambda(\mathrm{t}) \geq 0$ throughout $t \in [0, T]$.

Along with $\lambda(t) \geq 0$ and $\dot{\lambda}<0$, (23) implies there exists a time point t^* such that $\lambda(t^*) = \alpha$, $0 \leq t^* \leq T$, and thus $u^* = \begin{cases} a, 0 \leq t \leq t^* \\ 0, t^* < t \leq T \end{cases}$. According to (20), we can further show

$$x^* = \begin{cases} at + 1, 0 \leq t \leq t^* \\ at^* + 1, t^* < t \leq T \end{cases}. \tag{25}$$

We evaluate t^* by taking into account two conditions, $\lambda(t^*) = \alpha$ and $\lambda(T) = 0$. From (24) and (25) for $t^* \leq t \leq T$, $\dot{\lambda} = cmD(at^* + 1)^{m-1}$ and $\lambda = cmD(at^* + 1)^{m-1}t + k$. Since $\lambda(T) = 0$, $k = -cmD(at^* + 1)^{m-1}T$ and $\lambda = cmD(at^* + 1)^{m-1}(t - T)$.

Finally, $\lambda(t^*) = cmD(at^* + 1)^{m-1}(t^* - T) = \alpha$ and t^* must satisfy

$$(at^* + 1)^{m-1}(t^* - T) = \frac{\alpha}{cmD}. \tag{26}$$

We need to do numerical analysis to evaluate t^* satisfying (26). With t^*, the optimal costate, control, and state variables can be graphed as in Fig. 10.

For evaluating whether the supplier subsidy strategy makes the manufacturing firm better off, we need to compare the profit (J^*) based on the strategy with that involving no such supplier subsidy. Define J_0^* as the total profit the manufacturing firm can earn without supporting its supplier's innovation.

Theorem 1 *In order for the supplier subsidy strategy to be profitable for the manufacturing firm, it must be satisfied that* $J^* - J_0^* > 0$ *and therefore,*

$$D(c - \hat{c})T + \frac{D}{a(m+1)}\{c - (1 - amt^*)\hat{c}\} - \alpha at^* > 0$$

where $\hat{c} \equiv cx(t^*)^m = c(at^* + 1)^m$.

Proof See Appendix "Proof of Theorem 1".

However, the condition in Theorem 1 is just from the manufacturer's perspective. What makes the manufacturing firm better off might not necessarily do the same thing for the supplier. For a balanced analysis, we need to take the supplier's view as well.

Denote J^{s^*} as the supplier's total profit if the supplier accepts the manufacturer's subsidy, and $J_0^{s^*}$ as that if the supplier does not accept the deal.

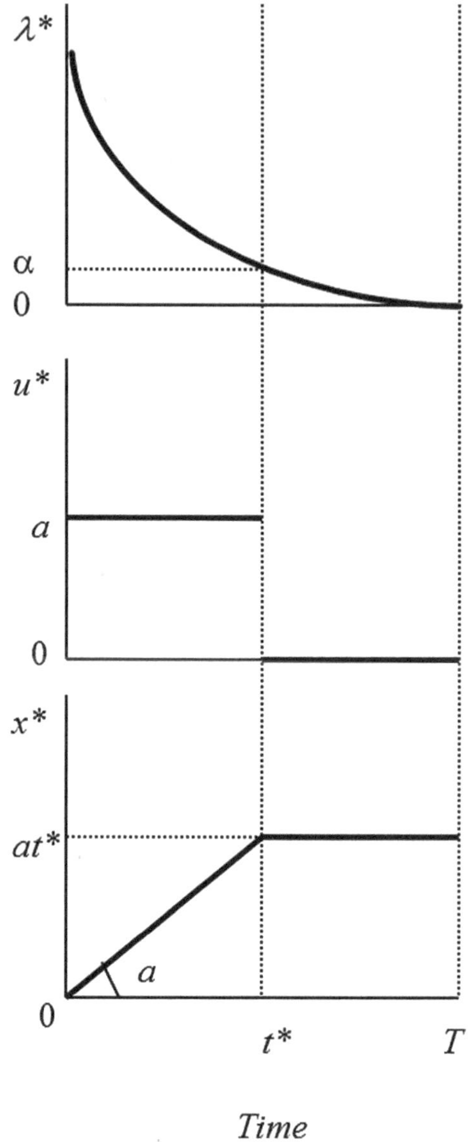

Fig. 10 A graphical interpretation of the optimal control solutions

Theorem 2 *The manufacturing firm's supplier subsidy strategy is beneficial to the supplier only if*

$$J^{s^*} - J_0^{s^*} = \beta a t^* - D(c - \hat{c})T - \frac{D}{a(m+1)}\{c - \hat{c}(1 - amt^*)\} > 0,$$

where β *is defined as the net project benefit the supplier can obtain from the manufacturer's support for a unit project.*

Proof See Appendix "Proof of Theorem 2".

What is more important is the following theorem based on the previous two theorems.

Theorem 3 *Suppose that the manufacturer's net outlay for the supplier subsidy per unit project,* α*, is not less than the supplier's net benefit,* β*, and that the production capacity (thus, the market demand) D is constant regardless of the supplier innovation, the manufacturing firm's supplier subsidy strategy cannot be profitable to both the manufacturer and its supplier concurrently. If* $\alpha = \beta$*, the manufacturer's profit increase due to the supplier innovation is exactly the same as the amount of loss experienced by the supplier due to the same innovation.*

Proof From Theorem 1, the subsidy strategy is profitable to the manufacturer only if $D(c-\hat{c})T + \frac{D}{a(m+1)}\{c-(1-amt^*)\hat{c}\} > \alpha at^*$. Also, from Theorem 2, the strategy can be acceptable to the supplier only if $\beta at^* > D(c-\hat{c})T - \frac{D}{a(m+1)}\{c-\hat{c}(1-amt^*)\} > 0$. Therefore, the subsidy strategy would be adopted by both manufacturer and supplier only if the two inequality relationship are satisfied, i.e., $\beta at^* > D(c-\hat{c})T - \frac{D}{a(m+1)}\{c-\hat{c}(1-amt^*)\} > \alpha at^*$. That is, a necessary condition must be met that $\beta at^* > \alpha at^*$, i.e., $\beta > \alpha$. However, according to our earlier argument about the 'realistic condition' in Appendix "Proof of Theorem 2", $\beta < \alpha$ holds rather than $\beta > \alpha$. As a result, we can state that in general, the subsidy strategy cannot be acceptable to both manufacturer and supplier simultaneously if the market demand for the final product does not change as the product price varies.

Moreover, if $\alpha = \beta$, from Theorems 1 and 2, it can be shown that $J^{s^*} - J_0^{s^*} = -(J^* - J_0^*)$, which proves the theorem.

In this section, we have proved that a manufacturing firm's simple strategy to support its supplier's innovation that can lead to the supply cost reduction is not enough to guarantee increased profitability to both the manufacturer and its supplier simultaneously. That is, in order for the supply chain to have an increased profit as a whole, it is not sufficient to make an improvement limited within the supply chain itself alone.

In the next section, we take a look at the possibility that the innovation subsidy strategy can benefit both the manufacturing firm and its supplier. In particular, we concentrate on the condition under which such strategy can make the two participants better off by considering a market demand structure sensitive to the price change of the final product.

2.1.2 An Extended Model: The Case of Price-Dependent Market Demand

As mentioned before, in this extended model, we assume that the market demand for the manufacturing firm's product is a function of the product price set by the manufacturer. More specifically, we adopt a simple linear demand curve well established in the literature: $D = d_1 - d_2 p$, where d_1 is the base (e.g., theoretical maximum) demand for the product and d_2 measures a demand sensitivity in response to the price change, i.e., $\frac{dD}{dp} = -d_2$, which represents the marginal change in demand as the price changes by a unit.

Further, we assume that the manufacturing firm tries to keep its unit net profit constant when it allows the product price to be reduced in order to increase the market demand. This assumption is not unusual: Sterman et al. (1997) showed that a firm's markup ratio remained remarkably constant, while the product cost and its market price decreased together. In the following model, the price is determined by the manufacturing firm as follows: $p = r + c_1 + cx^m$, where r is the fixed markup.

Then, $D(t) = d_1 - d_2(r + c_1 + cx^m)$.

It is practical to assume $D(t) = d_1 - d_2(r + c_1 + cx^m) > 0$ throughout the ensuing analysis: Therefore, we focus on d_2 that makes $D(t) > 0$ valid.

Now the manufacturing firm faces the following optimization problem,

*P*4:

$$Maximize\ J = \int_0^T [r(d_1 - d_2 r - d_2 c_1 - d_2 cx^m) - \alpha u] \mathrm{d}t$$

$$Subject\ to\ \dot{x} = u$$

$$0 \le u \le a, x(0) = x_0 = 1.$$

As in solving *P*3, we obtain the new Hamiltonian, $H = r(d_1 - d_2 r - d_2 c_1 - d_2 cx^m) - \alpha u + \lambda u$, and optimal solutions for *P*4 as follows.

Applying the same reasoning as in the simple model case, we can derive t^* which should satisfy $(at^* + 1)^{m-1}(t^* - T) = \frac{\alpha}{cmrd_2}$ so that $\begin{cases} x^* = at + 1, \text{while}\, t \le t^* \\ x^* = at^* + 1, \text{while}\, t > t^* \end{cases}$.

Denote $\tilde{J}^*$ as the total profit the manufacturer can earn by subsidizing the supplier innovation when its market demand is a linear function of the product price, and $\tilde{J}_0^*$ as the total profit the manufacturer can get without such subsidy.

Theorem 4 *When the market demand is a linear function of the product price, in particular,* $D(t) = d_1 - d_2(r + c_1 + cx^m)$, *the supplier subsidy strategy is beneficial to the manufacturing firm if and only if* $\tilde{J}^* - \tilde{J}_0^* > 0$, *thus*

$$d_2 r\left\{c\left(1+\frac{1}{a(m+1)T}\right)-\hat{c}\left(1+\frac{1-amt^*}{a(m+1)T}\right)\right\}T > \alpha a t^* \quad \text{or} \qquad (27)$$
$$d_2 r(C_A - C_B)T > \alpha a t^*,$$

where $C_A = c\left(1+\frac{1}{a(m+1)T}\right)$, $C_B = \hat{c}\left(1+\frac{1-amt^*}{a(m+1)T}\right)$ *and* $\hat{c} \equiv cx(t^*)^m = c(at^*+1)^m$.

Proof See Appendix "Proof of Theorem 4".

We need to examine the implication of Theorem 4 a little further. In order for the manufacturer to subsidize the supplier innovation, the resulting profit increase should be at least the same as the amount of subsidy spent for the innovation projects, $\alpha a t^*$. From (27), the condition mentioned above can be expressed as $d_2 r(C_A - C_B)T = \alpha a t^*$. That is, $d_2 r(C_A - C_B)T$ can be regarded as the profit increase due to the subsidy strategy for the supplier innovation, which can be further decomposed by rearranging terms:

(a) if we suppose that $C_A - C_B$ is the average cost saving (and therefore, price reduction) at each $t \in [0, T]$ because of the supplier innovation, from the manufacturer's point of view,
(b) $d_2(C_A - C_B)$ is the average demand increase due to the supplier innovation per unit period, and
(c) $d_2 r(C_A - C_B)$ is the profit increase per unit period, and finally, we can conclude
(d) $d_2 r(C_A - C_B)T$ is the manufacturer's total revenue increase over $t \in [0, T]$ resulting from the supplier subsidy strategy.

As in the simple model analysis, the strategy should benefit the supplier if it can be practically implemented. Denote $\tilde{J}^{s^*}$ as the supplier's optimal total profit when it accepts the manufacturer's subsidy for the innovation, and $\tilde{J}_0^{s^*}$ as that without involving any such subsidy.

Theorem 5 *In order for the supplier subsidy strategy under the variant market demand structure to be acceptable to the supplier, it must be satisfied that*

$$\beta a t^* > d_1\{\hat{c}t^* + S_F T - S_E\} - d_2\{\hat{c}t^* S_B + S_A S_F T - S_C S_E - S_D\}, \qquad (28)$$

where $S_A = r + c_1 + c + \hat{c} - c_s$, $S_B = r + c_1 + \hat{c} - c_s$, $S_C = r + c_1 - c_s$, $S_D = \frac{1}{a(2m+1)}[\hat{c}^2(at^*+1) - c^2]$, $S_E = \frac{1}{a(m+1)}[\hat{c}(at^*+1) - c]$, $S_F = c - \hat{c}$.

Proof See Appendix "Proof of Theorem 5".

It is not easy to understand the practical implications of S_A to S_E. But, we can try to make an indirect inference of the meaning of (28) by rearranging the terms so that $\{\hat{c}t^* + S_F T - S_E\}\{d_1 - d_2\frac{\hat{c}t^* S_B + S_A S_F T - S_C S_E - S_D}{\hat{c}t^* + S_F T - S_E}\}$, where $\{d_1 - d_2\frac{\hat{c}t^* S_B + S_A S_F T - S_C S_E - S_D}{\hat{c}t^* + S_F T - S_E}\}$ can be regarded as the average demand at $t \in [0, T]$ while $\{\hat{c}t^* + S_F T - S_E\}$ as the supplier's revenue loss per unit demand for $t \in [0, T]$. Thus, $\{\hat{c}t^* + S_F T - S_E\}\{d_1 - d_2\frac{\hat{c}t^* S_B + S_A S_F T - S_C S_E - S_D}{\hat{c}t^* + S_F T - S_E}\}$ represents the total revenue loss to the supplier for $t \in [0, T]$ by accepting the manufacturer's subsidy for supply innovation. As a result, Theorem 5 states that the supplier subsidy strategy is acceptable to the supplier if and only if the total net subsidy from the manufacturer should be larger than the total revenue loss to the supplier.

Theorem 6 *Given the conditions assumed in this section, the manufacturing firm's strategy to support its supplier innovation can be acceptable to both the manufacturer and the supplier if and only if* (27) *and* (28) *are satisfied simultaneously. That is,*

$$d_2 r \left\{ c\left(1 + \frac{1}{a(m+1)T}\right) - \hat{c}\left(1 + \frac{1 - amt^*}{a(m+1)T}\right) \right\} T > \alpha a t^*$$

and

$$\beta a t^* > d_1\{\hat{c}t^* + S_F T - S_E\} - d_2\{\hat{c}t^* S_B + S_A S_F T - S_C S_E - S_D\}.$$

As in the constant demand situation, if $\alpha > \beta$ is held, the condition becomes

$$d_2 r \left\{ c\left(1 + \frac{1}{a(m+1)T}\right) - \hat{c}\left(1 + \frac{1 - amt^*}{a(m+1)T}\right) \right\} T > \alpha a t^* > \beta a t^* > d_1\{\hat{c}t^* + S_F T - S_E\} - d_2\{\hat{c}t^* S_B + S_A S_F T - S_C S_E - S_D\} \tag{29}$$

Unlike Theorem 3, Theorem 6 indicates that it could be possible for both the manufacturing firm and its supplier to become better off from manufacturer's supporting the supplier innovation. Under what circumstances (29) holds depends on many factors such as characteristics of the market demand as expressed in d_1 and d_2, decision time horizon captured by T, and the supplier innovation capability embedded in m, i.e., its learning rate. However, the most critical difference between the model assumptions is concerned with the nature of the market demand structure. Therefore, in the numerical examples, it is reasonable for us to explore the impact of market demand structure on the profitability of supplier innovation support.

2.2 Numerical Examples

To draw more managerial insights from Theorem 6, we present numerical examples. In this section, we report key numerical results using parameter values in Table 7. Since we are interested in how the market sensitivity affects the supplier innovation support strategy, we focus on the numerical examples in which (29) is satisfied by varying d_2.

Numerical examples are depicted in Figs. 11, 12, 13, 14, and 15: Since we are concerned with $D > 0$, we focus on d_2 such that $0 < d_2 < 50$, which covers practically meaningful ranges. Figure 11 compares the manufacturing firm's total profit

Table 7 Parameter values for numerical analysis

T	a	c	c_1	c_s	d_1	ϕ_m	χ	r	α	β
100	10	4	4	1	500	0.95	2	2	5	5

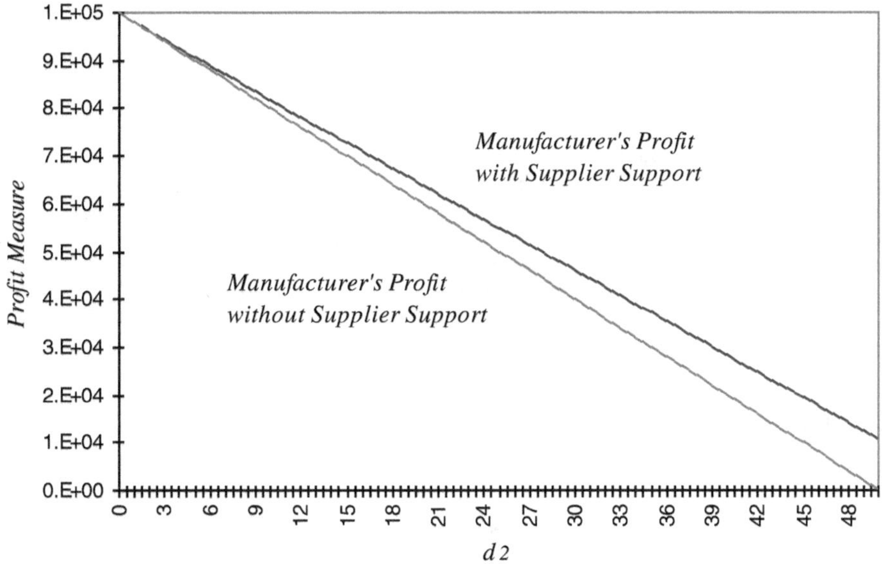

Fig. 11 Manufacturing firm's profit comparison

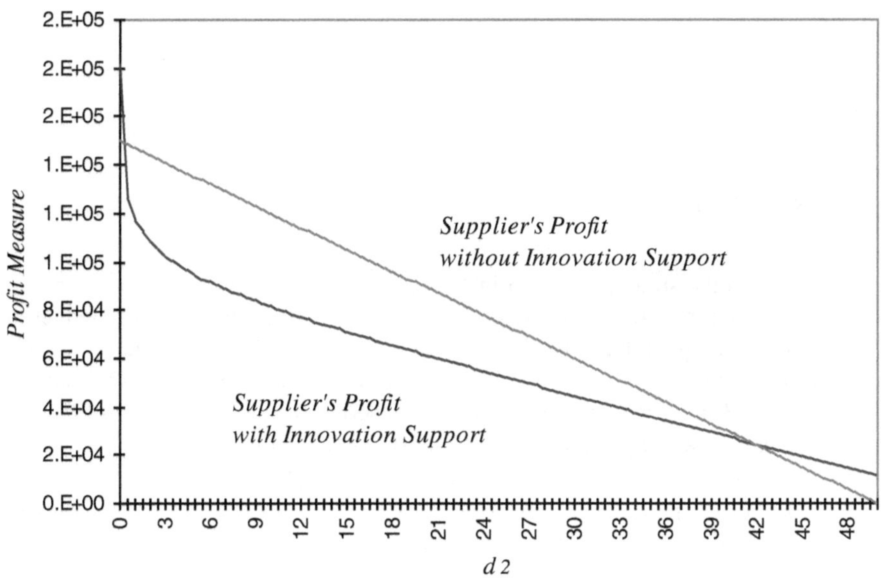

Fig. 12 Supplier's profit comparison

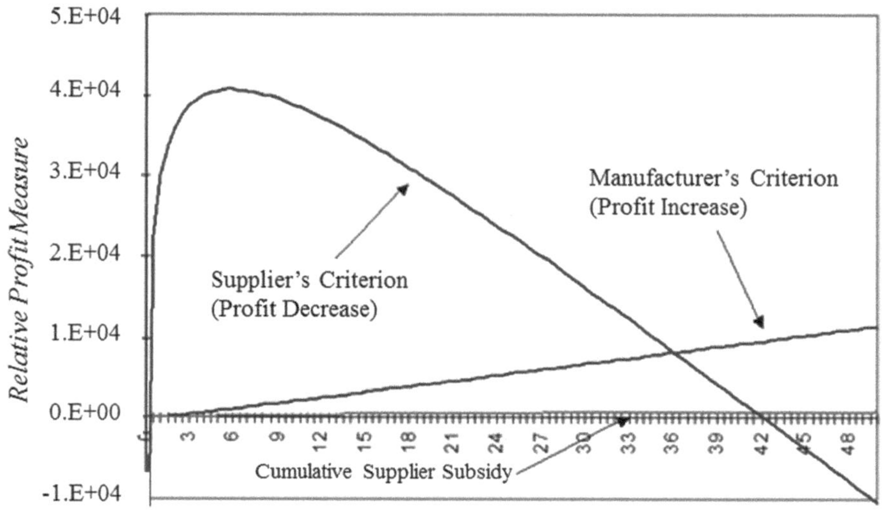

Fig. 13 Determining the joint positive profit range

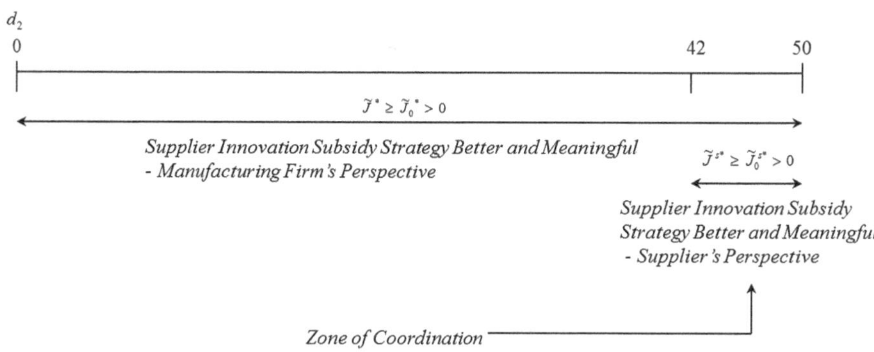

Fig. 14 Range for mutually beneficial coordination

associated with the strategy to coordinate, i.e., support, its supplier innovation with that not adopting the strategy as the market demand sensitivity, d_2, varies, depicting (83) and (84) in Appendix "Proof of Theorem 4". Figure 12 shows the same comparison for the supplier, graphing (85) and (86) in Appendix "Proof of Theorem 5". Finally, Fig. 13 integrates the two figures to describe the relationship in (29).

From Fig. 11, we can see that overall the manufacturing firm gets better off by supporting the supplier innovation regardless of the demand sensitivity, i.e., $\tilde{J}^* \geq \tilde{J}_0^*$. Although the difference between two strategies, one with supplier subsidy $(\tilde{J}^*)$ and the other without it $(\tilde{J}_0^*)$, is negligible when d_2 is very small (in particular, $d_2 \leq 5.5$), after the initial period, $\tilde{J}^*$ becomes larger than $\tilde{J}_0^*$.

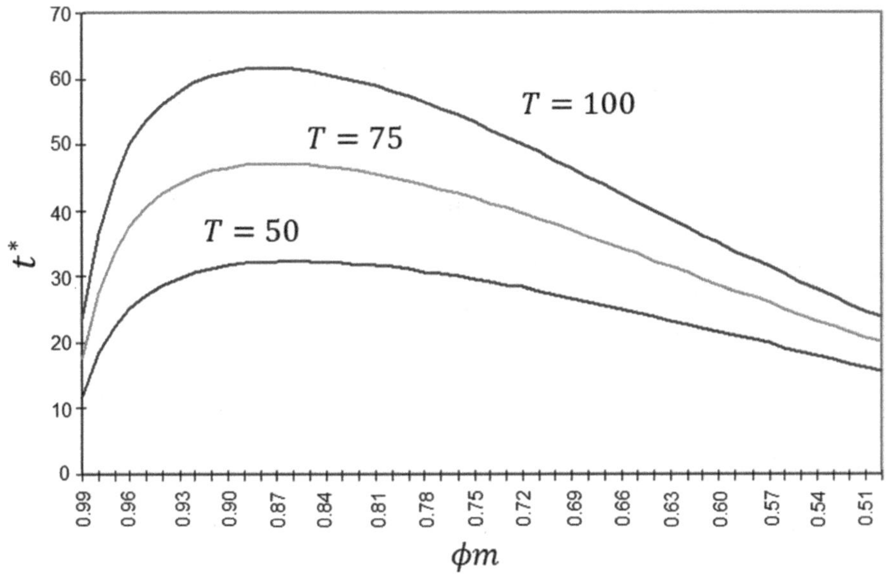

Fig. 15 Changes in t^* as ϕ_m varies, given $d_2 = 43$

For the supplier, we can draw a similar conclusion from Fig. 12: $\tilde{J}_0^{s^*}$ is larger than $\tilde{J}^{s^*}$ until d_2 becomes about 42.0, after which it holds $\tilde{J}^{s^*} > \tilde{J}_0^{s^*}$. Thus, we can infer that there is a nonlinear relationship between $\tilde{J}^{s^*}$ and $\tilde{J}_0^{s^*}$ for the supplier case.

Conclusions derived from Figs. 11 and 12 can be combined in Fig. 13, where at about $d_2 = 42.0$, the coordination strategy becomes acceptable to the supplier as well. For both the manufacturer and the supplier, the range of $42.0 \leq d_2 < 50.0$ makes it mutually beneficial to coordinate the supplier innovation that can eventually reduce the supply cost and attract more market demand. In effect, $42.0 \leq d_2 < 50.0$ can be called 'zone of coordination' (see Fig. 14).

From Fig. 13, we can make two practical observations. First, it is very large d_2 that can make the subsidy strategy acceptable to the supplier. This implies that the initial demand for the product is very small since $D = d_1 - d_2 p$. Second, a large d_2 also means that the demand increase as the product price decreases is very large: A small reduction in the supply cost should have a significant impact on the demand increase.

What kind of industry can most benefit from the supplier–manufacturer coordination like this? Based on the two observations above, we can infer that it might be an industry in between introduction and growing stages in the product life cycle that can benefit most: in this kind of industry, we can expect a relatively small initial market demand, but a huge potential for demand growth as innovation occurs.

The principal implication of this numerical exercise is to show that it is possible for both manufacturer and supplier to get better off by coordinating the supplier innovation *if* the market demand for the final product is sensitive to the product price set by the manufacturing firm to a certain extent. Despite being relied on numerical examples, this is a quite strong conclusion since Theorem 3 has mathematically proved that this kind of mutual benefit cannot be expected *if* the market demand is constant regardless of the supplier innovation. An important managerial implication is that the supply chain coordination cannot be profitable unless the efficiency improvement inside the supply chain can induce more revenues from outside the chain, i.e., the market.

Since our interest is concerned with the effect of market demand sensitivity on the supplier–manufacturer coordination, we have so far examined the manufacturer's and its supplier's profit changes as d_2 varies. Utilizing the similar approaches along with the analytical solution in the previous sections, we can analyze other dynamics involving different parameters. In Fig. 15, we examine how the supplier's innovation capability (captured in its learning rate) affects t^* (the optimal time length of innovation support) for three different Ts: Although $d_2 = 43$ is used for this particular numerical example, one can see a similar pattern for any reasonable value of d_2.

From Fig. 15, we can infer as follows. When the supplier innovation capability is relatively low, e.g., for about $\phi_m > 0.96$, t^* is relatively short: The reason could be that the innovation capability is too low to warrant a reasonable return for the innovation subsidy by the manufacturer. Likewise, if the capability is very high, e.g., for about $\phi_m < 0.65$, t^* is relatively short: In this case, the capability is so high that even a small amount of subsidy can contribute a lot, and therefore, it is not necessary to support many innovation projects. Finally, when the supplier's innovation capability is moderate, e.g., about $0.80 < \phi_m < 0.93$, the need for innovation support is the highest, implying that the coordination of innovation is most effective in that range. Another point we can make is that as the decision horizon, T, increases, so does t^*: The longer the decision horizon, the more endurable the effect of innovation support.

We can use the analysis model to investigate other important dynamics associated with different parameters as well as variables.

2.3 Conclusion and Managerial Implications

Our primary inquiry in this section has been to identify conditions associated with market demand that can make the supplier–manufacturer coordination for the supplier innovation mutually profitable to both participants. We have shown that if the market is not sensitive to the product price, and thus eventually the supplier innovation itself, then the supplier–manufacturer coordination cannot satisfy both the manufacturing firm and the supplier simultaneously. The extended analysis along with numerical examples has indicated that it could be possible for such supplier innovation subsidy to be acceptable to both participants at the same time if the market demand increases in response to the favorable innovation coordinated by the manufacturer and its supplier. An important implication is that unless the revenue increases outside of the supply chain itself, any efficiency gains within the supply chain cannot be sustainable since they are traded off between the chain participants, i.e., it is a zero-sum game if there is no substantive revenue increase from the outside. An analogy might be that as long as the absolute size of a pie remains fixed for two persons, it is never possible to allocate larger pieces to the two people simultaneously than those they got before.

Exercise Problems

1. Consider a supply chain in which a manufacturing firm produces a product and a supplier provides a key component to the manufacturer. A market demand is fixed as D, and a sales price is fixed as p. It is assumed that p is large enough for the manufacturer to have a positive profit. The manufacturer buys the key component from the supplier at price $c_s + m$, where $c_s(t)$ is the supplier's unit production cost at time t and m is the supplier's unit constant margin. The manufacturer's unit production cost is $c_m(t)$ at time t. Cost dynamics for the supplier are $\dot{c}_s = f(u)$ and $c_s(0) = c_{s0}$, and those for the manufacturer are $\dot{c}_m = g(v)$ and $c_m(0) = c_{m0}$, where $u(t)$ is the supplier's process innovation investment rate and $v(t)$ is the manufacturer's process innovation investment rate. For both the manufacturer and the supplier, the process innovation investment reduces the unit production cost with diminishing returns, i.e., $\frac{\partial f}{\partial u} < 0$, $\frac{\partial^2 f}{\partial u^2} > 0$, $\frac{\partial g}{\partial v} < 0$, and $\frac{\partial^2 g}{\partial v^2} > 0$. Innovation investment costs u^2 to the supplier and v^2 to the manufacturer. Also, the maximum innovation investment rate are $\bar{u}$ and $\bar{v}$, i.e., $u \leq \bar{u}$ and $v \leq \bar{v}$. The planning period is $t \in [0, T]$ and the discounting rate is r.
   (a) Formulate the optimal control problem for the coercive decision-making case, i.e., the manufacturer has a much larger bargaining power than the supplier so that the manufacturer exerts its power to determine innovation investment rates u and v that maximize its own profit.
   (b) Formulate the optimal control problem for the balanced decision-making case, i.e., the manufacturer and the supplier cooperatively determine the innovation investment rates u and v that maximize the supply chain profit (i.e., sum of the manufacturer's profit and the supplier's profit).
2. Consider a supply chain with a downstream manufacturer and an upstream part supplier. Product quality Q is the sum of Q_1 and Q_2, where the assembly quality Q_1 increases with quality investment rate q_1 of the manufacturer (i.e., $\dot{Q}_1(t) = q_1(t), Q_1(0) = Q_{10}$) and the part quality Q_2 increases with quality investment rate q_2 of the supplier (i.e., $\dot{Q}_2(t) = q_2(t), Q_2(0) = Q_{20}$). Product quality Q influences the market demand D via $D = \alpha - \beta p + \gamma Q$, where α is a market potential, and p is a sales price. Quality investment is costly, i.e., it costs $c_1 q_1^2$ to the manufacturer and $c_2 q_2^2$ to the supplier, and due to the resource limits, the maximum quality investment rates are $\bar{q}_1$ to the manufacturer and $\bar{q}_2$ to the supplier, respectively. Further, the manufacturer has a much larger bargaining power

than the supplier and decides the optimal investment rate $q_1(t)$ as well as $q_2(t)$ in order to maximize its profit over the planning period $0 \le t \le T$. We assume a short enough T so that the discounting effect can be disregarded and the sales price p and the wholesale price w are exogenous $(p > w)$.

(a) Derive the equilibrium investment strategies.
(b) The investment strategies in (a) concerns the manufacturer's incentives only. In which condition(s), would the investment strategies be feasible from the supplier's perspective?

3. Suppose there are two firms that produce substitutable products for the same market. Assume that a product price p is given for both firms and each firm can differentiate its product by improving its service reputation Q_i. Specifically, the consumer demand D_i for firm $i\ (i = 1, 2)$ is determined as follows: $D_i = \alpha - \beta p + \gamma(Q_i - Q_{3-i})$. Each firm can increase its service reputation Q_i by investing $u_i(t)$ in its service infrastructure for the customers, i.e., $\dot{Q}_i = u_i(t)$. Investing $u_i(t)$ incurs an investment cost of $\frac{ku_i^2}{2}$ and the initial service reputation of each firm is Q_0.

(a) Find the optimal $u_i(t)$ and $Q_i(t)$ over t $\in$ [0, T].
(b) Suppose that the two firms decide to cooperate. Therefore, they now jointly decide $u_i(t)$ to maximize the sum of their profits. Find the optimal path of $Q_i(t)$ and compare the result with that in (a). Explain the differences, if any.

Appendix 1

A Single Player-Dominated (Coercive) Case

We have the following Hamiltonian or Lagrangian:

$$\begin{aligned} L &= H \\ &= (x_1 + x_2)(a_1 - (x_1 + x_2)) - \rho u_2^2 + \lambda_1 u_1 + \lambda_2 u_2 + w_1(\bar{u}_1 - u_1) + w_2(\bar{u}_2 - u_2) \end{aligned} \tag{30}$$

The set of necessary conditions is as follows:

$$L_{u_1} = -2\rho u_1 + \lambda_1 - w_1 = 0 \tag{31}$$

$$L_{u_2} = \lambda_2 - w_2 = 0 \tag{32}$$

$$\dot{\lambda}_1 = -L_{x_1} = -(a_1 - (x_1 + x_2) - (x_1 + x_2)) = -(a_1 - 2(x_1 + x_2)) \\ = 2(x_1 + x_2) - a_1 \tag{33}$$

$$\dot{\lambda}_2 = -L_{x_2 =} - (a_1 - (x_1 + x_2) - (x_1 + x_2)) = 2(x_1 + x_2) - a_1 \tag{34}$$

$$w_1 \geq 0, w_1(\bar{u}_1 - u_1) = 0 \tag{35}$$

$$w_2 \geq 0, w_2(\bar{u}_2 - u_2) = 0 \tag{36}$$

- Behavior of λ_i

If $(x_1 + x_2) < \frac{a_1}{2}$, $\dot{\lambda}_1 < 0$ and $\dot{\lambda}_2 < 0$. Likewise, if $(x_1 + x_2) > \frac{a_1}{2}$, $\dot{\lambda}_1 > 0$ and $\dot{\lambda}_2 > 0$. Since $\dot{x}_1 = u_1 \geq 0$ and $\dot{x}_2 = u_2 \geq 0$, $\dot{\lambda}_1 < 0 \rightarrow \dot{\lambda}_1 > 0$ and $\dot{\lambda}_2 < 0 \rightarrow \dot{\lambda}_2 > 0$ as time increases. Moreover, $\lambda_1(T) = \lambda_2(T) = 0$.

Therefore, we reach the following conclusions.

(a) $$\dot{\lambda}_1 < 0 \text{ and } \dot{\lambda}_2 < 0 \quad \text{for } 0 \leq t \leq T \tag{37}$$

(b) Since $\dot{\lambda}_1 = \dot{\lambda}_2$ and $\lambda_1(T) = \lambda_2(T) = 0, \lambda_1 = \lambda_2$ throughout $0 \leq t \leq T$. (38)

(c) Therefore, from (32), $\lambda_1 = \lambda_2 = w_2$. (39)

(d) Since $w_2 > 0$ for $0 \leq t < T, u_2^* = \bar{u}_2$ throughout $0 \leq t < T$. (40)

- Behavior of w_1

Now, from (31), $\lambda_1 = 2\rho u_1 + w_1 \geq 0$.

(a) If $\lambda_1(0) < 2\rho\bar{u}_1$, then $w_1 = 0$ and $u_1^* < \bar{u}_1$ for $0 \leq t \leq T$. (41)

(b) But, if there exists $\hat{t}$ such that

$$\lambda_1(t) = 2\rho\bar{u}_1 + w_1(t) \text{ for } 0 \leq t \leq \hat{t} < T \text{ and } w_1(\hat{t}) = 0 \text{ i.e., } \lambda_1(0) > 2\rho\bar{u}_1, \\ \text{then } u_1^* = \bar{u}_1 \text{ for } 0 \leq t \leq \hat{t} \text{ and } u_1^* < \bar{u}_1 \text{ for } \hat{t} \leq t \leq T. \tag{42}$$

- How to determine $\hat{t}$?

First, suppose $w_1 > 0$ and thus $u_1^* = \bar{u}_1$. From (33), $\dot{\lambda}_1 = 2(x_1 + x_2) - a_1$.

$$\ddot{\lambda}_1 = 2(\dot{x}_1 + \dot{x}_2) = 2(\bar{u}_1 + \bar{u}_2), \tag{43}$$

$$\dot{\lambda}_1 = 2(\bar{u}_1 + \bar{u}_2)t + k_1, \tag{44}$$

$$\lambda_1 = (\bar{u}_1 + \bar{u}_2)t^2 + k_1 t + k_2, \tag{45}$$

where k_1 and k_2 are constants to be determined.

Now, from (31) and $w_1 > 0$, we have

$$w_1 = \lambda_1 - 2\rho\bar{u}_1 = (\bar{u}_1 + \bar{u}_2)t^2 + k_1 t + k_2 - 2\rho\bar{u}_1 \tag{46}$$

(46) indicates that $w_1 \geq 0$ is a quadratic function, and we can find the minimum of w_1 by arranging (46) as follows:

$$w_1 = (\bar{u}_1 + \bar{u}_2)\left\{t + \frac{k_1}{2(\bar{u}_1 + \bar{u}_2)}\right\}^2 - \frac{k_1^2}{4(\bar{u}_1 + \bar{u}_2)} + k_2 - 2\rho\bar{u}_1. \tag{47}$$

Therefore, in order for $\hat{t} > 0$ in (42) to exist, it must hold that

$$-\frac{k_1^2}{4(\bar{u}_1 + \bar{u}_2)} + k_2 - 2\rho\bar{u}_1 = 0 \tag{48}$$

and

$$\hat{t} = \frac{-k_1}{2(\bar{u}_1 + \bar{u}_2)} > 0. \tag{49}$$

(48) implies $k_1^2 - 4(\bar{u}_1 + \bar{u}_2)(k_2 - 2\rho\bar{u}_1) = 0$ and thus

$$\mathrm{k}_1 = \pm 2\sqrt{(\bar{u}_1 + \bar{u}_2)(k_2 - 2\rho\bar{u}_1)}. \tag{50}$$

From (45), we know that $\mathrm{k}_2 = \lambda_1(0)$.

Plugging (50) into (49) appropriately, we finally have

$$\hat{t} = \sqrt{\frac{\lambda_1(0) - 2\rho\bar{u}_1}{\bar{u}_1 + \bar{u}_2}}. \tag{51}$$

A Balanced Decision-Making Case

We have the following Hamiltonian or Lagrangian:

$$\begin{aligned} L &= H \\ &= (x_1 + x_2)(a_1 + a_2 - 2(x_1 + x_2)) - \rho_2 u_1^2 \\ &\quad - \rho_2 u_2^2 + \lambda_1 u_1 + \lambda_2 u_2 + w_1(\bar{u}_1 - u_1) + w_2(\bar{u}_2 - u_2) \end{aligned} \tag{52}$$

The set of necessary conditions is as follows:

$$L_{u_1} = -2\rho_1 u_1 + \lambda_1 - w_1 = 0 \tag{53}$$

$$L_{u_2} = -2\rho_2 u_2 + \lambda_2 - w_2 = 0 \tag{54}$$

$$\dot{\lambda}_1 = -(a_1 + a_2 - 2(x_1 + x_2) - 2(x_1 + x_2)) = 4(x_1 + x_2) - a_1 - a_2 \tag{55}$$

$$\dot{\lambda}_2 = -(a_1 + a_2 - 2(x_1 + x_2) - 2(x_1 + x_2)) = 4(x_1 + x_2) - a_1 - a_2 \tag{56}$$

$$w_1 \geq 0, w_1(\bar{u}_1 - u_1) = 0 \tag{57}$$

$$w_2 \geq 0, w_2(\bar{u}_2 - u_2) = 0, \tag{58}$$

- Behavior of λ_i

From (55) and (56), $\dot{\lambda}_1 = \dot{\lambda}_2 < 0$. Using the same reasoning as in Appendix 1 along with $\lambda_1(T) = \lambda_2(T) = 0$, we conclude:

(a) $\lambda_1 = \lambda_2 \geq 0$ and λ_i is decreasing throughout

$$0 \leq t \leq T. \tag{59}$$

- Behavior of w_i

If $w_i > 0$, then $u_i = \bar{u}_i$ according to (57) and (58). From (53) and (54), we know that while

$$w_i > 0, w_i = \lambda_i - 2\rho_i \bar{u}_i \tag{60}$$

From (60), we know w_i and λ_i have the same slope while $w_i > 0$. That is,

$$\dot{w}_i = \dot{\lambda}_i \text{ while } w_i > 0. \tag{61}$$

Suppose w_i is for the player i, then we define w_{-i} for the other player. Therefore, we have: Using $\lambda_1 = \lambda_2 \equiv \lambda$,

$$w_i = \lambda_i - 2\rho_i \bar{u}_i = \lambda - 2\rho_i \bar{u}_i \text{ while } w_i > 0, \tag{62}$$

$$w_{-i} = \lambda_{-i} - 2\rho_{-i} \bar{u}_{-i} = \lambda - 2\rho_{-i} \bar{u}_{-i} \text{ while } w_{-i} > 0. \tag{63}$$

From (62) and (63), we know that if

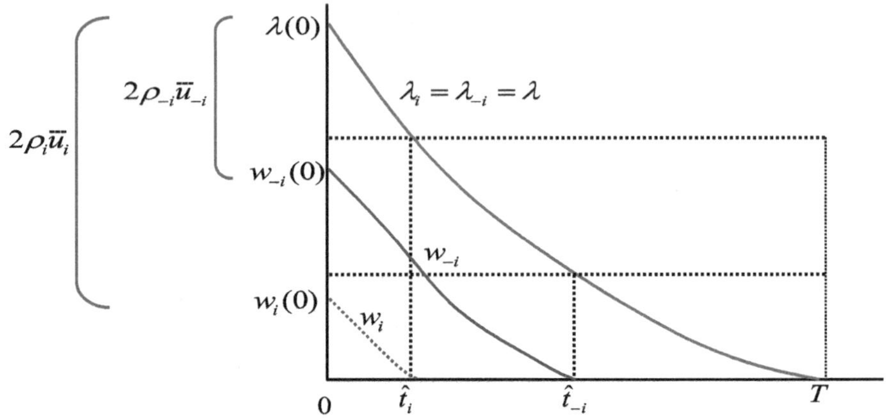

Fig. 16 Behaviors of w_i and w_{-i}

Table 8 Dynamic solutions for u_i^* and u_{-i}^*

Cases	0 – $\hat{t}_i$	$\hat{t}_i$ – $\hat{t}_{-i}$	$\hat{t}_{-i}$ – T
$\lambda(0) > 2\rho_i\bar{u}_i$	$u_i^* = \bar{u}_i$	$u_i^* < \bar{u}_i$	
		$u_{-i}^* = \bar{u}_{-i}$	$u_{-i}^* < \bar{u}_{-i}$
$2\rho_{-i}\bar{u}_i < \lambda(0) < 2\rho_i\bar{u}_i$		$u_i^* < \bar{u}_i$	
		$u_{-i}^* = \bar{u}_{-i}$	$u_{-i}^* < \bar{u}_{-i}$
$\lambda(0) < 2\rho_{-i}\bar{u}_{-i}$		$u_i^* < \bar{u}_i$	
		$u_{-i}^* < \bar{u}_{-i}$	

$$\rho_i\bar{u}_i > \rho_{-i}\bar{u}_{-i}, w_i < w_{-i} \tag{64}$$

Figure 16 depicts the situation specified in (64). Now suppose that $\hat{t}_i$ and $\hat{t}_{-i}$ exist such as in Fig. 16. Then, we can suggest the dynamics of u_i^* and u_{-i}^* at least in a figurative manner (see Table 8).

- How to determine $\hat{t}_i$?

In order to determine the specific $\hat{t}_i$ s, we assume $\rho_1\bar{u}_1 > \rho_2\bar{u}_2$ and thus $w_2 > w_1$, while

$$w_2 > w_1 > 0, u_1 = \bar{u}_1 \text{ and } u_2 = \bar{u}_2. \tag{65}$$

From (55), we derive the following:

$$\ddot{\lambda}_1 = 4(u_1 + u_2) = 4(\bar{u}_1 + \bar{u}_2),$$

$$\dot{\lambda}_1 = 4(\bar{u}_1 + \bar{u}_2)t + k_1$$

$$\lambda_1 = 2(\bar{u}_1 + \bar{u}_2)t^2 + k_1 t + k_2. \tag{66}$$

Now, from (62),

$$w_1 = \lambda_1 - 2\rho_1\bar{u}_1 = 2(\bar{u}_1 + \bar{u}_2)t^2 + k_1 t + k_2 - 2\rho_1\bar{u}_1. \tag{67}$$

By rearranging (67), we obtain:

$$w_1 = 2(\bar{u}_1 + \bar{u}_2)\left[t + \frac{k_1}{4(\bar{u}_1 + \bar{u}_2)}\right]^2 - \frac{k_1^2}{8(\bar{u}_1 + \bar{u}_2)} + k_2 - 2\rho_1\bar{u}_1 \tag{68}$$

We take the similar steps with (48)–(51).
From (68), if $\hat{t}_1 < T$ indeed exists such that $w_1(\hat{t}_1) = 0$, then

$$\hat{t}_1 = \frac{-k_1}{4(\bar{u}_1 + \bar{u}_2)} \text{ and } -\frac{k_1^2}{8(\bar{u}_1 + \bar{u}_2)} + k_2 - 2\rho_1\bar{u}_1 = 0. \tag{69}$$

From (69),

$$k_1 = \pm 2\sqrt{2(\bar{u}_1 + \bar{u}_2)(k_2 - 2\rho_1\bar{u}_1)}. \tag{70}$$

From (66),

$$k_2 = \lambda_1(0). \tag{71}$$

Using (69)–(71) appropriately, we obtain:

$$\hat{t}_1 = \sqrt{\frac{\lambda_1(0) - 2\rho_1\bar{u}_1}{2(\bar{u}_1 + \bar{u}_2)}}. \tag{72}$$

Now, we would like to determine $\hat{t}_2$.
From (65), since $\rho_1\bar{u}_1 > \rho_2\bar{u}_2$,

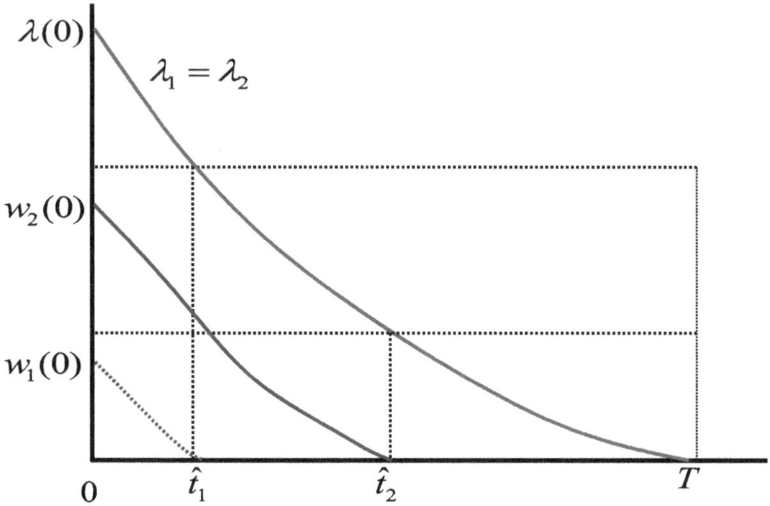

Fig. 17 Determining $\hat{t}_2$

$$\lambda(0) > 2\rho_1\bar{u}_1 > 2\rho_2\bar{u}_2. \tag{73}$$

If $\rho_1\bar{u}_1 > \rho_2\bar{u}_2$ and there exists $\hat{t}_1$ such that $\hat{t}_1 < T$, there must exist $\hat{t}_2$ such that

$$\hat{t}_1 < \hat{t}_2 < T. \tag{74}$$

As in (61), $\dot{w}_i = \dot{\lambda}_i$ while $w_i > 0$, implying that $w_1(t)$ and $w_2(t)$ have the same slope as that of $\lambda(t)$ as long as $w_i > 0$. Using this property along with the relationship in Fig. 17, we derive the following:

$$\frac{\lambda(0) - w_2(0)}{T - \hat{t}_2} = \frac{\lambda(0) - w_1(0)}{T - \hat{t}_1}. \tag{75}$$

From (60), we know

$$w_i(0) = \lambda(0) - 2\rho_i\bar{u}_i. \tag{76}$$

Therefore, by plugging (76) into (75), we obtain:

$$\hat{t}_2 = T - \frac{\rho_2\bar{u}_2}{\rho_1\bar{u}_1}(T - \hat{t}_1). \tag{77}$$

Appendix 2

Proof of Theorem 1

With (25), we can calculate the optimal profit the manufacturing firm can earn by employing the supplier support strategy. That is,

$$J^* = \int_0^{t^*} [D(p - c_1 - c(at+1)^m) - \alpha a]\mathrm{d}t + [D(p - c_1 - c(at^*+1)^m)](\mathrm{T} - \mathrm{t}^*). \tag{78}$$

The second term in (78) represents the profit for $t \in [t^*, T]$: Since $x^* = at^* + 1$ and $u^* = 0$ for $t > t^*$, $\int_{t^*}^{T} [D(p - c_1 - c(at^*+1)^m)]\mathrm{d}t = [D(p - c_1 - c(at^*+1)^m)](\mathrm{T} - \mathrm{t}^*)$.

Therefore, the total manufacturer's profit becomes

$$J^* = D(p - c_1)T - \hat{c}D(T - t^*) - \alpha a t^* + \frac{D}{a(m+1)}\{c - \hat{c}(at^*+1)\},$$

where $\hat{c} \equiv cx(t^*)^m = c(at^*+1)^m$.

In order to evaluate whether the supplier subsidy strategy makes the manufacturing firm better off, we need to compare the profit (J^*) based on the strategy with that involving no such supplier subsidy. Define J_0^* as the total profit the manufacturing firm can earn without supporting its supplier's innovation.

We can calculate

$$J_0^* = D(p - c_1 - c)T. \tag{79}$$

J^* can be rearranged so that

$$J^* = D(p - c_1 - c)T + D(c - \hat{c})T + \hat{c}Dt^* - \alpha a t^* + \frac{D}{a(m+1)}\{c - \hat{c}(at^*+1)\} \tag{80}$$

Therefore, in order for the subsidy strategy to be attractive to the manufacturing firm, it must be that $J^* > J_0^*$ or $J^* - J_0^* > 0$. By combining (79) and (80), we obtain the necessary condition for the profitable supplier subsidy strategy as

$$\begin{aligned} J^* - J_0^* &= D(c - \hat{c})T + \hat{c}Dt^* - \alpha a t^* + \frac{D}{a(m+1)}\{c - \hat{c}(at^*+1)\} \\ &= D(c - \hat{c})T + \frac{D}{a(m+1)}\{c - (1 - amt^*)\hat{c}\} - \alpha a t^* > 0 \end{aligned}$$

Proof of Theorem 2

We can calculate

$$J^{s^*} = \int_0^{t^*} [D(c(at+1)^m - c_s) + \beta a]\mathrm{d}t + \int_{t^*}^{T} D(\hat{c} - c_S)\mathrm{d}t \tag{81}$$

where c_S is the supplier's unit internal manufacturing cost: It can be either raw material cost or other base costs that cannot be reduced through the innovation.

Let us define β as the net project benefit the supplier can obtain from the manufacturer's support for a unit project or experiment. If the cost the supplier must spend in order to implement the project or experiment is $\gamma_s \geq 0$ (it is the external expenses, i.e., direct cost, the supplier must pay to conduct an innovation project), $\beta = \gamma_m - \gamma_s$. From the manufacturer's perspective, the total cost per project is α, while the supplier's net benefit from the support is only β, which remains within the supplier after the project has been done: β becomes an embedded asset to the supplier. In general, we can impose a realistic constraint that $\gamma_s < \gamma_m < \alpha$ and $\beta = \gamma_m - \gamma_s < \alpha - \gamma_s < \alpha$.

Equation (81) takes the particular form since the supplier's net profit from supplying products (intermediate goods) to the manufacturer is $D(c(at^* + 1)^m - c_s)$ at $t \leq t^*$, but $D(c(at^* + 1)^m - c_s)$ at t such that $t^* < t \leq T$.

Equation (81) can be rewritten as follows:

$$\begin{aligned} J^{s^*} &= D(\hat{c} - c_S)(T - t^*) + (\beta a - c_s D)t^* + \frac{D}{a(m+1)}(\hat{c}(at^* + 1) - c) \\ &= D(c - c_s)T + D(\hat{c} - c)T + (\beta a - \hat{c}D)t^* + \frac{D}{a(m+1)}(\hat{c}(at^* + 1) - c) \end{aligned}$$

On the other hand, if the deal is rejected by the supplier, the total profit would be

$$J_0^{s^*} = D(c - c_S)T. \tag{82}$$

Therefore, if the supplier subsidy strategy is beneficial to the supplier, it must be satisfied that $J^{s^*} - J_0^{s^*} > 0$. From (81) and (82), we have the following:

$$J^{s^*} - J_0^{s^*} = \beta a t^* - D(c - \hat{c})T - \frac{D}{a(m+1)}\{c - \hat{c}(1 - amt^*)\}.$$

Proof of Theorem 4

Each of the profits can be determined as follows:

$$\begin{aligned}\tilde{J}^* &= \int_0^{t^*} \{r(d_1 - d_2 r - d_2 c_1 - d_2 c(at+1)^m - \alpha a\}\mathrm{d}t + \int_{t^*}^{T} \{r(d_1 - d_2 r - d_2 c_1 - d_2 \hat{c})\}\mathrm{d}t \\ &= r(d_1 - d_2 r - d_2 c_1 - d_2 c)T + d_2 r(c - \hat{c})T + \frac{d_2 r}{a(m+1)}\{c - \hat{c}(1 - amt^*)\} - \alpha a t^*.\end{aligned} \tag{83}$$

$$\tilde{J}_0^* = r(d_1 - d_2(r + c_1 + c))T = r(d_1 - d_2 r - d_2 c_1 - d_2 c)T. \tag{84}$$

In order for the subsidy strategy to be beneficial to the manufacturing firm, it must be satisfied that $\tilde{J}^* > \tilde{J}_0^*$. From (83) and (84),

$$\begin{aligned}&\tilde{J}^* - \tilde{J}_0^* = d_2 r(c - \hat{c})T + \frac{d_2 r}{a(m+1)}\{c - \hat{c}(1 - amt^*)\} - \alpha a t^* > 0 \text{ implies} \\ &d_2 r\left\{c\left(1 + \frac{1}{a(m+1)T}\right) - \hat{c}(1 + \frac{1 - amt^*}{a(m+1)T})\right\}T > \alpha a t^*.\end{aligned}$$

Proof of Theorem 5

Applying the previous reasoning, we calculate

$$\begin{aligned}\tilde{J}^{s^*} &= \int_0^{t^*} \{[d_1 - d_2(r + c_1 + c(at+1)^m)](c(at+1)^m - c_s) + \beta a\}\mathrm{d}t \\ &+ \int_{t^*}^{T} \{[d_1 - d_2(r + c_1 + \hat{c})](\hat{c} - c_s)\}\mathrm{d}t\end{aligned} \tag{85}$$

and

$$\tilde{J}_0^{s^*} = [d_1 - d_2(r + c_1 + c)](c - c_s)T, \tag{86}$$

where c_s is the fixed, i.e., constant, portion of the supplier's unit internal manufacturing cost.

By simplifying and rearranging (85) and (86), we determine

$$\tilde{J}^{s^*} - \tilde{J}_0^{s^*} = \beta a t^* - d_1\{\hat{c}t^* + S_F T - S_E\} + d_2\{\hat{c}t^* S_B + S_A S_F T - S_C S_E - S_D\}$$

and therefore, $\tilde{J}^{s^*} - \tilde{J}_0^{s^*} > 0$ implies

$$\beta a t^* > d_l\{\hat{c}t^* + S_F T - S_E\} - d_2\{\hat{c}t^* S_B + S_A S_F T - S_C S_E - S_D\}$$

where $S_A = r + c_1 + c + \hat{c} - c_s$, $S_B = r + c_1 + \hat{c} - c_s$, $S_C = r + c_1 - c_s$, $S_D = \frac{1}{a(2m+1)}[\hat{c}^2(at^* + 1) - c^2]$, $S_E = \frac{1}{a(m+1)}[\hat{c}(at^* + 1) - c]$, $S_F = c - \hat{c}$.

Appendix 3: Discontinuous Control Variable

Consider an optimal control theory problem, which is linear in the control variable $u(t)$ as follows:

$$Max \int_{t_0}^{t_1} (f_1(t, x(t)) + u(t) f_2(t, x(t)))dt$$

$$\begin{aligned} Subject\ to\ x'(t) &= g_1(t, x(t)) + u(t) g_2(t, x(t)), \\ x(t_0) &= x_0, \\ b_l &\le u(t) \le b_u. \end{aligned}$$

To solve the problem, we could apply the Lagrangian (see Appendix 4 in Chap. 1). Since the particular structure of the problem makes it possible to find a simpler solution method, we structure the usual Hamiltonian H as follows:

$$H = f_1(t, x(t)) + u(t) f_2(t, x(t)) + \lambda(t) g_1(t, x(t)) + \lambda(t) u(t) g_2(t, x(t)).$$

The necessary conditions call for $\frac{\partial H}{\partial u} = f_2 + \lambda g_2 = 0$, which is of little use because it does not help us to obtain the optimal control variable u^*.

Note that the goal of this problem is to maximize the objective function, which is equivalent to maximizing H. Since $b_l \le u(t) \le b_u$, we have the following relationship so as to maximize the objective function:

$$u^* = \left\{ \begin{matrix} b_l \\ ? \\ b_u \end{matrix} \right\} \quad \text{if} \quad f_2 + \lambda g_2 \left\{ \begin{matrix} < \\ = \\ > \end{matrix} \right\} 0.$$

Chapter 3
Innovation Competition and Strategy

Abstract Innovation enables the firm to develop and improve strong competitive advantage in the market. However, innovation doesn't arise in a vacuum. Competition stirs innovation. That's why I focus on innovation competition in the context of value chain. Innovation competition is a crucial factor in operations strategy. In this chapter, I first show how a firm's learning capability affects its innovation competition and then explore the conditions under which competing firms are willing to collaborate to build a common infrastructure as public goods.

Keywords Innovation · Innovation competition · Value chain · Learning capability · Collaboration

Key Learning

- *How is the firm's innovation affected by its learning capability and the market's reward for the innovation?*
  - The firm with lower learning capability would invest less in innovation.
  - The faster the market reward depreciates as the development delays, the increasing rate of investment slows down more as time advances.
  - The more the market reward depreciates, the increasing rate of probability that the firm completes the innovation successfully decreases more.
- *Is it possible to avoid an opportunistic behavior when firms collaborate to build a common infrastructure, which has characteristics of public goods?*
  - In general, the free-riding issue is prevalent when building a common infrastructure.
  - When including both common and firm-specific benefits simultaneously in the model, we can show it is possible to avoid the free-riding problem in building a common infrastructure.

The original version of this chapter was revised. The erratum to this chapter is available at 10.1007/978-981-10-3599-9_6

B. Kim, *Optimal Control Applications for Operations Strategy*,
DOI 10.1007/978-981-10-3599-9_3

- If the firm-specific benefit is sufficiently large compared with the common benefit, it will be possible to avoid the free-riding problem even when there is no collusive arrangement between the firms.
- If the firm can find a way to enhance its own internal productivity, i.e., a firm-specific benefit, by utilizing the knowledge derived from the 'common effort,' it should get *more* involved in developing the common infrastructure.

Innovation enables the firm to sustain its competitive advantage. Since innovation must occur in a dynamic environment, however, the firm should manage the innovation process strategically. In general, firm's products or services embody its innovation capability. As such, new product or service development is an essential element of innovation management. Because of the competitive nature of innovation, differential games modeling can shed light on understanding the firm's dynamic innovation process as well as new product development by analyzing the complex interface between key players in the market. Therefore, we first discuss basic principles of dynamic programming and differential games: our objective here is not to teach the student complete theories for these subjects, but to present key elements to refresh the student's prior knowledge about these subjects. Then, we consider two examples of firm's dynamic innovation management. The first examines the firms' innovation competition, i.e., two firms are competing against each other in developing a new product, where firms' learning capabilities and profit structures are taken into account. The other looks into firms' cooperation as well as competition in building a common infrastructure. Comparing the infrastructure to a public good, we consider a business setting, where competing firms are willing to cooperate for building the infrastructure, but they are to compete against each other by capitalizing on the infrastructure once it is completed.

1 Basics of Dynamic Programming

In the previous chapters, we have utilized Pontryagin's maximum principle to solve optimal control theory problems. Here we consider another approach, Dynamic Programming, which we can use to solve optimal control theory problems. Consider an optimal control theory problem:

$$Max \int_{t_0}^{t_1} f(t, x(t), u(t))dt + \theta(x(t_1), t_1)$$

$$\begin{aligned} \textit{Subject to} \quad & x'(t) = g(t, x(t), u(t)) \\ & x(t_0) = x_0,\ t_0,\ t_1 \text{ fixed (given), and } x(t_1) \text{ free (unfixed)} \\ & \theta(x(t_1), t_1) \text{ is a salvage term.} \end{aligned}$$

Suppose we can define a value function $J(t,x(t))$ that has the following characteristics: $J(t_1,x(t_1)) = \theta(x(t_1),t_1)$ and $J(t_0,x_0) = max_u\left(\int_{t_0}^{t_0+\Delta t} fdt + \int_{t_0+\Delta t}^{t_1} fdt + \theta\right)$.

Then, we can find the Hamilton–Jacobi–Bellman (HJB) equation:

$$-J_t(t,x) = max_u(f(t,x(t),u(t)) + J_x(t,x)g(t,x(t),u(t))).$$

For an infinite horizon autonomous problem, we suggest a simpler form of HJB equation. Consider an infinite horizon autonomous problem as follows:

$$Max \int_0^{\infty} e^{-rt} f(x,u)dt$$

$$Subject\ to\ x'(t) = g(x,u)$$

Then, the revised HJB equation for optimality is: $rJ(x) = max_u(f(x,u) + J'(x)g(x,u))$.

After discoursing differential games, we explain how to use the Hamilton–Jacobi–Bellman equation to solve a differential games problem.

2 Basics of Differential Games

Until now, we have mainly focused on optimal control theory problems, where there is only one decision making entity. On the contrary, in the real world, it is prevalent that a decision-making problem involves multiple stakeholders. That is, in every market situation, more than one decision makers are competing with each other. Think about the smartphone market. There are multiple companies that make and sell smartphones in the global market. Two largest players are Apple and Samsung, which compete with each other to have a larger market share and get more profit. In this competitive situation, one company's performance is determined not only by its own strategy, but also by its competitor's. In effect, the two companies' strategies are dynamically interacting with each other. As such, in order to analyze one company's strategy, we have to consider the dynamic interaction between the two competitors simultaneously. Differential games enable us to accomplish the goal.

Consider a differential games problem, where there are two competitors, i.e., players. Each player tries to optimize her own objective function. The two players' strategies are intertwined by the constraints that are common for the two decision makers at the same time. That is, the common state variables are controlled by the same differential equations.

A general structure of differential games is presented as follows:

$$\text{for player 1, } \textit{Maximize}\, V_1(u_1,u_2) = \int_{t_0}^{t_1} f_1(t,x_1(t),x_2(t),u_1(t),u_2(t))dt$$

$$\text{for player 2, } \textit{Maximize}\, V_2(u_1,u_2) = \int_{t_0}^{t_1} f_2(t,x_1(t),x_2(t),u_1(t),u_2(t))dt$$

$$\textit{Subject to}\, x_i'(\mathrm{t}) = g_i(t,x_1(t),x_2(t),u_1(t),u_2(t))$$

$$x_i(t_0) = x_{i0},\ x_i(t_i)\ \text{free}, i = 1,2,$$

the common constraints for both players.

When solving a differential games problem, our goal is to find a Nash equilibrium, which implies that given its competitor's optimal control u_{3-i}^*, player i's optimal control u_i^* should make the value of player i's objective function larger than or equal to that with the nonoptimal control u_i.

More specifically, Nash equilibrium solutions mean $V_1(u_1^*,u_2^*) \geq V_1(u_1,u_2^*)$ for player 1 and $V_2(u_1^*,u_2^*) \geq V_2(u_1^*,u_2)$ for player 2.

There are two approaches to solving a differential games problem, i.e., open-loop and feedback strategies.

2.1 Open-Loop Solution

An open-loop solution implies that at the outset of the game, each player chooses her entire course of action, i.e., all the values of the control variables, throughout the decision time horizon. It implicitly assumes that the player is committed to the initial strategy and does not intend to revise it down the road. It is relatively straightforward to obtain an open-loop solution for a differential game problem.

First, for each player, we form a Hamiltonian as follows:

$$H_i(t,x_1(t),x_2(t),u_1(t),u_2(t),\lambda_{i1}(t),\lambda_{i2}(t)) = f_i + \lambda_{i1}g_1 + \lambda_{i2}g_2,\ i = 1,2$$

A Nash equilibrium for this approach means:

$$H_1(t,x_1,x_2,u_1^*,u_2^*,\lambda_{11},\lambda_{12}) \geq H_1(t,x_1,x_2,u_1,u_2^*,\lambda_{11},\lambda_{12}) \text{ and}$$
$$H_2(t,x_1,x_2,u_1^*,u_2^*,\lambda_{21},\lambda_{22}) \geq H_2(t,x_1,x_2,u_1^*,u_2,\lambda_{21},\lambda_{22}).$$

Finally, the necessary conditions are the same with those for the Pontryagin's maximum principle. That is, necessary conditions are:

$$H_{iu_i} = \frac{\partial H_i}{\partial u_i} = 0,\ i = 1,2 \text{ and } \lambda_{ij}' = -\frac{\partial H_i}{\partial x_j},\ i = 1,2, j = 1,2.$$

2.2 Feedback Solution

Although the open-loop solution is relatively easy to understand and derive, it might not be realistic enough. That is, in the real world, the decision maker constantly monitors the environment, which is reflected in the state variables of the differential games, and adjusts and refines her strategies, i.e., the control variables, so as to take into account the environmental context optimally. A feedback solution is consistent with the reality exemplified above. The feedback strategy is the one, where the player continuously revises her strategy as the states evolve. In effect, the player's control variable is formulated as $u_i = u_i(t, x_1(t), x_2(t))$, $i = 1, 2$ to highlight the nature of the feedback solution to revise the player's strategy as the state variables develop further down the road. The biggest conceptual difference between the open-loop and the feedback strategies is that the latter is more compatible with the contingency strategy, which clearly advocates that the best choice is contingent on the current state of the system, i.e., the dynamic game.

Considering the dynamic nature of the feedback solution, we correctly expect that the feedback solution is more difficult to obtain. For the feedback solution, we define the Hamiltonian as follow:

$$\begin{aligned}&H_i(t, x_1(t), x_2(t), u_1(t, x_1(t), x_2(t)), u_2(t, x_1(t), x_2(t)), \lambda_{i1}(t), \lambda_{i2}(t))\\&= f_i + \lambda_{i1} g_1 + \lambda_{i2} g_2, i = 1, 2.\end{aligned}$$

We write the necessary conditions:

$$\begin{aligned}H_{iu_i} &= \frac{\partial H_i}{\partial u_i} = 0,\ i = 1, 2\\ \lambda_{1j}^{'} &= -\frac{\partial H_1}{\partial x_j} - \left(\frac{\partial H_1}{\partial u_2}\right)\left(\frac{\partial u_2^*}{\partial x_j}\right), j = 1, 2\\ \lambda_{2j}^{'} &= -\frac{\partial H_2}{\partial x_j} - \left(\frac{\partial H_2}{\partial u_1}\right)\left(\frac{\partial u_1^*}{\partial x_j}\right), j = 1, 2.\end{aligned}$$

The way to calculate $\lambda_{ij}^{'}$ is completely different from that for an open-loop solution. In fact, it is the structure of $\lambda_{ij}^{'}$ that distinguishes the feedback from the open-loop strategy. Because of the complexity of the feedback solution, it is often necessary to apply the Hamilton–Jacobi–Bellman (HJB) equation to solving a differential games problem for the feedback strategy. That is, $-J_t(t, x) = max_u(f(t, x(t), u(t)) + J_x(t, x)g(t, x(t), u(t)))$ is commonly employed to find feedback strategies. See an example in Appendix 1.

3 Innovation Competition[1]

Firms' long-term competitiveness depends on their ability to carry out product or process innovation, which enables them to appropriate extraordinary rents for a certain period of time, say until another innovation makes it obsolete. In this section, we investigate a dynamic interplay between firms' learning capabilities, market reward structures for the innovation, and other cost factors, establishing an optimal control model for an innovation game context.

The analysis result enables us to construct a causal relationship map among the key decision elements that identifies a dynamic path through which a firm's learning capability affects its competitor's innovation strategy. Numerical examples based on the control model indicate that the firm with lower learning capability would invest less in innovation than would its competitor with higher learning capability. The faster the market reward to the innovation depreciates as the development delays, the more slows down the rate of investment increase as time advances. Likewise, the more the market reward depreciates, the more attenuates the increase rate of probability that the firm completes the innovation successfully.

3.1 A Continuous Dynamic Model

We establish an optimal control theory model, where firms' learning capabilities are explicitly incorporated into the analysis. The learning capabilities affect innovation cost structures that in turn impact the firms' innovation strategies, and thus their success probabilities. In addition to the learning capabilities, we expect the market reward for the innovation to depreciate over time. How fast the market reward depreciates influences the dynamics of the firm's innovation strategy.

3.1.1 Model Formulation

We consider an innovation game setting where there are two competitive players, i.e., Firm i, $i = 1, 2$. The two firms are competing with each other in innovating a new product (e.g., it can be a new process). In order to make the innovation successful, each firm has to acquire a certain level of scientific knowledge, which follows a probability distribution. That is, the firms do not know a priori exactly how much knowledge is required to succeed in innovating the product. We denote $x_i(t)$ as Firm i's knowledge level at time t, and $F_i(x_i(t))$ as the probability that Firm

[1]This section is a refined version of Kim B. (1999) "Dynamic Effects of Learning Capabilities and Profit Structures on the Innovation Competition." *Optimal Control Applications and Methods, 20*, 127–144.

i succeeds in innovating by t. Consistent with the decision-theoretic model assumptions, we can work with the following probability distribution.

$$F_i(x_i(t)) = 1 - e^{-\lambda x_i(t)} \text{ and} f_i(x_i(t)) \equiv \frac{dF_i(x_i)}{dx_i} = \lambda e^{-\lambda x_i}, \tag{1}$$

i.e., $f_i(x_i(t))$ is the probability density function for the success time distribution with regard to $x_i(t)$. Disregarding t in $x_i(t)$ for the moment, we can see that F_i is the cumulative distribution function of an exponential random variable with parameter λ. The most important attribute of an exponential distribution is the 'memoryless' property, which can be interpreted in the current context as follows. Denote $P_s(\tilde{T} > x_a)$ as the probability that it takes more than x_a to complete the innovation, i.e., the innovation has not been completed for $0 \leq x_i \leq x_a$. Then, according to the memoryless property, $P_s(\tilde{T} > x_a + x_b | \tilde{T} > x_a) = P_s(\tilde{T} > x_b)$: that is, how much it will take to complete the innovation from a particular knowledge level is independent of the particular knowledge level itself accumulated so far. This trait of the exponential distribution has been used to describe the game-theoretic situation in which the memoryless property as well as randomness to a certain extent can be observed extensively in a research-intensive environment (Canbolat et al. 2012).

Still disregarding t in $x_i(t)$, we denote $E(x_i)$ as the average level of knowledge that enables the firm to complete the innovation. Since we already know that x_i follows an exponential distribution with parameter λ from (1), $E(x_i) = 1/\lambda$. That is, the parameter λ is related with the average amount of knowledge required for succeeding in the innovation. It seems logical to assume that the firms competing for the innovation in the same industry must face the same industrial infrastructure that could determine the value of λ. Therefore, we use the same λ for both competitors.

We denote $u_i(t)$ as Firm i's effort devoted to the innovation at t, i.e., the control variable of the dynamic model. One way to understand a more realistic implication of $u_i(t)$ is to regard it as representing the number of projects or experiments conducted to enhance the firm's knowledge relevant to the innovation, ignoring the integer constraint. Then, there involves some form of coefficient that can transform the knowledge project or experiment to the knowledge stock.

Firm i's knowledge level accumulates according to the following dynamics:

$$\dot{x}_i(t) = \alpha_i u_i(t), \tag{2}$$

where $x_i(0) = 0$ and $u_i(t) \geq 0$ are assumed without loss of generality, and α_i is a coefficient to capture the efficiency of Firm i's ability to transform its innovation effort into knowledge necessary for the innovation. Since $x_i(t)$ is the state variable of the model, $x_i(0) = 0$ could imply that the current setting is congruent with an innovation game involving new product or process development for which each firm has very little prior experience or knowledge.

Firms' innovation efforts incur costs in which we incorporate firms' learning capabilities: we assume a firm's learning capability enables the firm to reduce its innovation cost as it accumulates more experience in innovating; the larger the learning capability, the faster the cost reduction occurs. Most of the learning curve models in manufacturing postulated that the learning occurs as cumulative production quantity increases. The assumption is based on a premise that manufacturing experience correlates with the cumulative production quantity. In this model, we associate the experience factor with time, i.e., the cost of innovation effort decreases as time passes according to the firm's learning rate. Under the situation where firms' technological experiences correlate with time progression, the assumption can be valid: in a competitive innovation environment, rapid technological improvement might be distributed evenly over a certain period time, and therefore, the technological experience might as well correlate with time.

Adopting the well-established learning function (Yelle 1979) with a time variable replacing the cumulative production quantity, we have the unit cost of Firm *i*'s innovation effort at *t*,

$$C_i(t) = K_i t^{n_i}, \tag{3}$$

where K_i is constant, $n_i = ln\phi_i/ln2$, and ϕ_i Firm *i*'s learning rate, $0 \leq \phi_i \leq 1$. In effect, K_i is the unit innovation cost at the very beginning of the innovation process, i.e., at $t = 1$. The implication of ϕ_i is that as time doubles, the unit innovation cost is reduced by $(1 - \phi_i) \times 100\%$: for instance, if $C_i(5) = 10$ and $\phi_i = 0.8$, then $C_i(10) = 8$.

As alluded already, the two competitors are in the same industry, and thus assumed to encounter the same industrial infrastructure that can be influential in inducing autonomous innovation. Moreover, since the two players try to do their best in the innovation game, their experiences in innovation efforts could be qualitatively equivalent. We have also posited that the two firms start their innovation efforts almost at the same time due to that the innovation involves a new product or process in which they have little prior experience. These observations allow us to use time (t) as approximating the cumulative amount of experience in innovation faced by both firms: as such, it is assumed that the two firms' autonomous experiences in the innovation are equivalent.

We assume that Firm *i*'s total cost be of a quadratic form depending on the amount of innovation effort at *t*. More specifically, presuming a quadratic cost function, we model the total cost as

$$TC_i(t, u_i) = K_i t^{n_i} u_i^2/2, \tag{4}$$

i.e., the total innovation cost is proportional to the unit cost in terms of $u_i^2/2$, where $u_i^2/2$ is used instead of u_i^2 because we want to simplify $\partial TC_i/\partial u_i$ without loss of generality.

From (3) and (4), we know $\partial TC_i/\partial t = n_i TC_i/t \leq 0$ and $\partial TC_i/\partial u_i = C_i u_i \geq 0$. In addition, $\partial^2 TC_i/\partial t^2 \geq 0$ and $\partial^2 TC_i/\partial u_i^2 \geq 0$. Thus, $TC_i(t, u_i)$ is decreasing concave with regard to *t* and increasing concave with regard to u_i. The total cost will be discounted according to the market discount rate *r*, which also represents the firm's opportunity cost for its innovation effort. Figure 1 shows example dynamics of total cost and unit cost based on the parameter values that will be later used for numerical analysis as in Table 1 except for the firms' learning capabilities so as to underscore the contrasting dynamics between the two firms' cost patterns: for the current example, we use $\phi_1 = 0.95$ and $\phi_2 = 0.80$ rather than $\phi_1 = 1.00$ and $\phi_2 = 0.90$ employed for the numerical examples.

Finally, we need a market reward function for which we make a particular assumption that the market recognizes the value of the innovation instantly when it is first completed, but perceives it less valuable as its completion time delays: this assumption is consistent with a type of innovation for which there is already a known need for the product. Under this kind of circumstance, the market may immediately reward the innovation, thus producing a decreasing reward function. Although one can express these characteristics by using the market discount rate, we opt not to do so since the market reward structure might behave in a way different from that the firm's opportunity cost does. Suppose the estimated market size for the innovation is R_0, in terms of the same (monetary) unit as that for the innovation cost, if the innovation is achieved at the instant of the beginning of the innovation process i.e., at $t = 1$. We can view R_0 as the present value of a lump-sum payment the market is willing to make for the successful innovation: in effect, this innovation game is like a patent competition, where only the first inventor takes all the reward, and therefore, the time elapse between the first successful innovation and the second is irrelevant to the game result (Canbolat et al. 2012).

As the completion time delays, however, the market will be less willing to pay the full reward for the innovation possibly because other products (or processes) alternative to or substitutes for the innovation in point might become available as time advances. This process of devaluation can take a different form from the market discount rate. Regarding R_0 as the constant lump-sum payment the market is willing to pay for the innovation if no alternative to or substitute for the innovation is available, we model the market reward at t as,

$$R(t) = R_0 t^m, \tag{5}$$

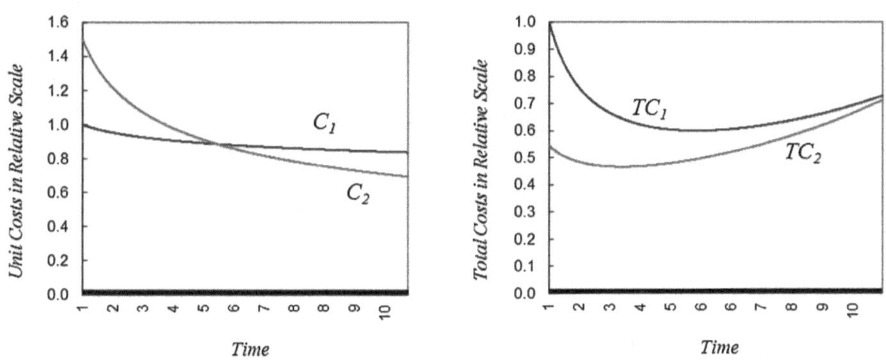

Fig. 1 Example dynamics of total cost and unit cost

Table 1 Summary of decision variables and elements

• $u_i(t)$: amount of Firm i's innovation effort at t, i = 1, 2, control variable,
• $x_i(t)$: stock of Firm i's technological knowledge for the innovation at t, state variable
• $\dot{x}_i(t) = \alpha_i u_i(t)$: dynamic equation for $x_i(t)$, where α_i is an efficiency coefficient,
• $F_i(x_i(t)) = 1 - e^{-\lambda x_i(t)}$: cumulative probability function that Firm i completes the innovation by t,
• $f_i(x_i(t)) \equiv \frac{dF_i(x_i)}{dx_i} = \lambda e^{-\lambda x_i}$: probability density function,
• $C_i(t) = K_i t^{n_i}$: Firm i's unit innovation cost, where K_i is constant, $n_i = ln\phi_i/ln2$, ϕ_i Firm i's learning rate,
• $TC_i(t, u_i) = K_i t^{n_i} u_i^2/2$: Firm i's total innovation cost at t,
• $R(t) = R_0 t^m$: market reward for the innovation if the innovation is completed at t, $m = ln\phi_m/ln2$, ϕ_m market reward depreciation rate,
• T: end of the decision time horizon, $[1, T]$

where $m = ln\phi_m/ln2$, ϕ_m, $0 \leq \phi_m \leq 1$, having an implication similar with that of ϕ_i. We further assume that after time T, the innovation will be of little value to the market, and therefore the current decision horizon is $[1, T]$, i.e., after T, the market does not value the innovation at all. If we want to be mathematically more rigorous, we have to impose that $R(t^+) = 0$ for $t^+ = T + \varepsilon$, where $\varepsilon > 0$. But, for the purpose of our current analysis, we do not need the additional constraint.

Note the decision time horizon for our analysis is $[1, T]$. By using $t = 1$ instead of $t = 0$ as the starting time for the decision horizon, we can avoid a technical difficulty when dealing with numerical examples. The reason is as follows. Since we use particular forms of cost and reward structures, e.g., $C_i(t) = K_i t^{n_i}$ and $R(t) = R_0 t^m$, we might have to encounter irregularity while calculating the cost and reward for $t \in [0, 1]$. It is purely a technical issue that does not affect the result of our analysis. Thus, without loss of generality, we use the decision horizon $[1, T]$ for our analysis as well as numerical examples. Table 1 summarizes some of the key decision elements in the model.

Given the formulation developed so far, we propose that Firm i's objective is to maximize

$$J^i(u_1, u_2) = \int_0^T \left[R(t)\left(1 - F_j(x_j)\right)\frac{dF_i(x_i(t))}{dt} - e^{-rt}(1 - F_1(x_1))(1 - F_2(x_2))TC_i(t, u_i)\right]dt,$$
$$i \neq j,\ i,j = 1, 2. \tag{6}$$

For Firm 1, i.e., $i = 1$, the objective function becomes

$$J^1(u_1, u_2) = \int_0^T \left[R(t)(1 - F_2(x_2))\frac{dF_1(x_1(t))}{dt} - e^{-rt}(1 - F_1(x_1))(1 - F_2(x_2))C_1(t)\frac{u_1^2}{2}\right]dt, \text{ where} \tag{7}$$

- $R(t)$ is the market reward for the innovation at t,
- $1 - F_2(x_2(t))$ is the probability that Firm 2 has not completed the innovation by t,
- $dF_1(x_1(t))/dt$ is the instantaneous probability that Firm 1 completes the innovation at t,
- $(1 - F_1)(1 - F_2)$ is the probability that neither firm has succeeded by t,
- $TC_1(t, u_1) = C_1(t)u_1^2/2$ is Firm 1's total innovation cost at t, and
- e^{-rt} is the discounting factor, where r is the market discounting rate.

We can see that the integrand of (7) consists of two parts: (i) the first part is the revenue Firm 1 would get if it succeeds in completing the innovation at t while Firm 2 has not been successful, i.e., $R(t)(1 - F_2(x_2))\frac{dF_1(x_1(t))}{dt}$, and (ii) the second part is the innovation cost the firm would have to pay at t if neither firm has been able to complete the innovation by t. Therefore, the firm tries to maximize the sum of the expected net innovation profit, i.e., 'innovation revenue—innovation cost,' over the entire decision time horizon, $[1, T]$.

We can see $\int_0^T R(t)(1 - F_2(x_2))\frac{dF_1(x_1(t))}{dt}dt$ in (7) can be rewritten as

$$\int_0^T R(t)(1 - F_2(x_2))\frac{dF_1(x_1(t))}{dt}dt = [R(t)(1 - F_2(x_2))F_1(x_1)]_0^T - \int_0^T F_1(x_1)d(R(t)(1 - F_2(x_2))) = \int_0^T F_1(R\dot{F}_2 - \dot{R}(1 - F_2))dt + R(T)(1 - F_2(x_2(T)))F_1(x_1(T)). \tag{8}$$

Taking into account the constraints, the decision problem faced by Firm 1 is to

$$Maximize\, J^1(u_1, u_2) = \int_0^T \left[RF_1\dot{F}_2 - \dot{R}F_1(1 - F_2) - e^{-rt}(1 - F_1)(1 - F_2)TC_1(t, u_1)\right]dt + R_0T^m\left(e^{-\lambda x_2(T)} - e^{-\lambda(x_1(T) + x_2(T))}\right)$$

$$\begin{aligned} Subject\ to\ \dot{x}_1(\mathrm{t}) &= \alpha_1 u_1(t) \\ \dot{x}_2(\mathrm{t}) &= \alpha_2 u_2(t). \end{aligned} \tag{9}$$

The constraints in (9) are from (2) and based on the premise that Firm 1 has to take into account not only its own knowledge dynamics constraint but also its competitor's reflecting the game-theoretic nature of the innovation.

Likewise, we can formulate a similar optimization problem for Firm 2 by replacing subscript 1 with subscript 2 and subscript 2 with subscript 1 in (9). Thus,

Firm 2 faces the same set of constraints, and both firms have to take into account the same market reward structure.

3.1.2 Analysis Results

Using a dynamic programming approach, we develop the HJB equation for (9) such as

$$-J_t^1(t, x_1, x_2) = \underset{u_1}{\text{Max}}\left[RF_1\dot{F}_2 - \dot{R}F_1(1-F_2) - e^{-rt}(1-F_1)(1-F_2)TC_1 + J_{x_1}^1\alpha_1 u_1 + J_{x_2}^1\alpha_2\bar{u}_2\right], \tag{10}$$

where $J_t^1 = \partial J^1/\partial t$ and $J_{x_i}^1 = \partial J^1/\partial x_i$. Instead of u_2, $\bar{u}_2$ is used to highlight that $\bar{u}_2$, i.e., Firm 2's strategy, is given to Firm 1 at each point of time when Firm 1 has to decide its own 'closed-loop or feedback' strategy. This does not imply that u_1 and u_2 are always independent: in fact, as shown later, each of them affects the other through a complex causal relationship. But, whenever each firm has to make a decision, the other's strategy at that particular moment can be regarded as given: since we are dealing with a continuous time decision horizon, this premise should not cause any analytical problem.

From (1), we can derive

$$\frac{dF_i(x_i(t))}{dt} = \frac{dx_i(t)}{dt}\frac{dF_i(x_i(t))}{dx_i(t)} = \alpha_i u_i(t)\lambda e^{-\lambda x_i}. \tag{11}$$

Inserting (11) along with specific formulas for decision variables into (10), we have

$$\begin{aligned} -J_t^1(t, x_1, x_2) = \text{Max}_{u_1}\Big[&R_0 t^m\left(1-e^{-\lambda x_1}\right)\lambda e^{-\lambda x_2}\alpha_2\bar{u}_2 - mR_0 t^{m-1}\left(1-e^{-\lambda x_1}\right)e^{-\lambda x_2} \\ &-K_1 t^{n_1}e^{-rt}e^{-\lambda(x_1+x_2)}\frac{u_1^2}{2} + J_{x_1}^1\alpha_1 u_1 + J_{x_2}^1\alpha_2\bar{u}_2\Big]. \end{aligned} \tag{12}$$

From (12), we can derive an optimal u_1, that is u_1^*, by taking a partial differentiation of $-J_t^1$ with regard to u_1 and setting it to zero:

$$\frac{d\left(-J_t^1\right)}{du_1} = -K_1 t^{n_1}e^{-rt}e^{-\lambda(x_1+x_2)}u_1 + \alpha_1 J_{x_1}^1 = 0. \tag{13}$$

Therefore,

$$\therefore u_1^* = \left[\frac{\alpha_1}{K_1} J_{x_1}^1 t^{-n_1} e^{rt} e^{\lambda(x_1+x_2)}\right]. \quad (14)$$

Likewise, for Firm 2,

$$\therefore u_2^* = \left[\frac{\alpha_2}{K_2} J_{x_2}^2 t^{-n_2} e^{rt} e^{\lambda(x_1+x_2)}\right]. \quad (15)$$

By putting (14) and (15) into the right-hand-side of (12) and rearranging it, we have

$$\begin{aligned}
&R_0 t^m \left(1-e^{-\lambda x_1}\right)\lambda e^{-\lambda x_2}\alpha_2\left[\frac{\alpha_2}{K_2} J_{x_2}^2 t^{-n_2} e^{rt} e^{\lambda(x_1+x_2)}\right] - mR_0 t^{m-1}\left(1-e^{-\lambda x_1}\right)e^{-\lambda x_2} \\
&-\frac{1}{2}K_1 t^{n_1} e^{-rt} e^{-\lambda(x_1+x_2)}\left[\frac{\alpha_1}{K_1} J_{x_1}^1 t^{-n_1} e^{rt} e^{\lambda(x_1+x_2)}\right]^2 + J_{x_1}^1\alpha_1\left[\frac{\alpha_1}{K_1} J_{x_1}^1 t^{-n_1} e^{rt} e^{\lambda(x_1+x_2)}\right] \\
&+J_{x_2}^1\alpha_2\left[\frac{\alpha_2}{K_2} J_{x_2}^2 t^{-n_2} e^{rt} e^{\lambda(x_1+x_2)}\right] = -mR_0 t^{m-1}\left(1-e^{-\lambda x_1}\right)e^{-\lambda x_2} + \left(J_{x_1}^1\right)^2 \frac{3\alpha_1^2}{2K_1} t^{-n_1} e^{rt} e^{\lambda(x_1+x_2)} \\
&+J_{x_2}^1 J_{x_2}^2 \frac{\alpha_2^2}{K_2} t^{-n_2} e^{rt} e^{\lambda(x_1+x_2)} + J_{x_2}^2 \frac{\alpha_2^2}{K_2} R_0 t^{m-n_2} e^{rt}\lambda\left(e^{\lambda x_1}-1\right).
\end{aligned} \quad (16)$$

Therefore, the entire equation of (12) can be rearranged as

$$\begin{aligned}
0 = J_t^1 &+ \left(J_{x_1}^1\right)^2 \frac{3\alpha_1^2}{2K_1} t^{-n_1} e^{rt} e^{\lambda(x_1+x_2)} + J_{x_2}^1 J_{x_2}^2 \frac{\alpha_2^2}{K_2} t^{-n_2} e^{rt} e^{\lambda(x_1+x_2)} \\
&+ J_{x_2}^2 \frac{\alpha_2^2}{K_2} R_0 t^{m-n_2} e^{rt}\lambda\left(e^{\lambda x_1}-1\right) - mR_0 t^{m-1}\left(1-e^{-\lambda x_1}\right)e^{-\lambda x_2}.
\end{aligned} \quad (17)$$

From (9) and the dynamic formulation in (10), we know at $t = T, J^i(T, x_1(T), x_2(T)) = R_0 T^m \left(e^{-\lambda x_j(T)} - e^{-\lambda(x_1(T)+x_2(T))}\right)$, which is the boundary condition for this optimal dynamic control problem.

In order to have a continuous optimal solution which is also in line with the boundary condition,

$$\begin{aligned}
&J^i(T, x_1(T), x_2(T)) = R_0 T^m \left(e^{-\lambda x_j(T)} - e^{-\lambda(x_1(T)+x_2(T))}\right), \text{ we propose} \\
&J^i(t, x_1, x_2) = h_i(t) e^{-\lambda(x_1+x_2)} + l_i(t) e^{-\lambda x_j}, i \neq j, i, j = 1, 2.
\end{aligned} \quad (18)$$

From (18), we can derive

$$\begin{aligned} J^i_{x_i} &= -\lambda h_i(t)e^{-\lambda(x_1+x_2)}, \\ J^i_{x_j} &= -\lambda h_i(t)e^{-\lambda(x_1+x_2)} - \lambda l_i(t)e^{-\lambda x_j}, \\ J^i_t &= \dot{h}_i(t)e^{-\lambda(x_1+x_2)} + \dot{l}_i(t)e^{-\lambda x_j},\ and\, h_1(T) = h_2(T). \end{aligned} \tag{19}$$

From the boundary condition and (18), we must have $h_i(T) = R_0T^m$, and therefore it is easy to see $h_1(T) = h_2(T)$.

Theorem 1 (The structure of an optimal solution) *The optimal solution for Firm 1 can be rewritten as* $u_1^* = \left[\frac{-\alpha_1\lambda h_1(t)}{C_1(t)e^{-rt}}\right]$. *In addition, the optimal solution requires that* $E_1(t) = 0$ *for* $h_1(t)$, *and the optimal* $l_1(t) = R_0t^m$,
where
$E_1(t) = \dot{h}_1 + \lambda^2 e^{rt}\left(\frac{3\alpha_1^2}{2K_1}t^{-n_1}h_1^2 + \frac{\alpha_2^2}{K_2}t^{-n_2}h_1h_2\right) + \lambda^2 e^{rt}\frac{\alpha_2^2}{K_2}R_0t^{m-n_2}h_2 + mR_0t^{m-1}.$

Proof From (3) and (19), it is easy to show $u_1^* = \left[\frac{-\alpha_1\lambda h_1(t)}{C_1(t)e^{-rt}}\right]$. For the second part of the theorem, see Appendix "Derivation of Theorem 1".

From Theorem 1, we can infer that the lesser is the present value of unit innovation cost $C_1(t)e^{-rt}$, the more innovation effort is made by the firm. The role played by $h_1(t)$ in determining u_1^* can be more clearly understood after we consider the competitive interaction between the two firms. Briefly speaking, the absolute value of $h_1(t)$, i.e., $-h_1(t)$ since $h_1(t)$ turns out to be always negative, represents the relative effectiveness of Firm 1's technological knowledge in increasing the firm's expected profit, compared with its competitor's: since we have already assumed $u_1^* \geq 0$, in order for $\left[\frac{-\alpha_1\lambda h_1(t)}{C_1(t)e^{-rt}}\right]$ to be positive, it must always hold that $h_1(t) < 0$.

As already mentioned, α_1 represents the efficiency of Firm 1's ability to transform its innovation effort into knowledge necessary for innovation. Finally, $1/\lambda$ is the average number of knowledge units, i.e., x_1, required for completing the innovation: as λ increases, the probability of completing the innovation increases given the same amount of technological knowledge. Thus, λ somehow measures an efficiency of the firm's technological knowledge in determining the success probability.

In what follows, we can schematically describe cause-and-effect relations of the key components in u_1^*:

- α_1, representing *knowledge transformation capability* which transforms innovation effort (u_i) into technological knowledge (x_i),
- $h_1(t)$, representing *profitability factor* that links technological knowledge (x_i) with profitability (J^i),
- λ, representing *success probability factor* that associates technological knowledge (x_i) with innovation success probability (F_i),
- $C_2(t)e^{-rt}$, denoting *innovation cost factor* which is the present value of innovation cost.

Thus, the firm's optimal innovation strategy implies

$$Optimal\_Innovation\_Effort \approx \frac{[Knowledge\ Transformation] \times [Profitability\ Factor] \times [Success\ Probability\ Factor]}{Innovation\ Cost\ Factor}.$$

Corollary 1 *For Firm 2, we derive similar results. That is,* $u_2^* = \left[\frac{-\alpha_2 \lambda h_2(t)}{C_2(t)e^{-rt}}\right]$ *which requires that* $G_1(t) = 0$ *for* $h_2(t)$, *and the optimal* $l_2(t) = R_0 t^m$, *where* $G_1(t) = \dot{h}_2 + \lambda^2 e^{rt}\left(\frac{3\alpha_2^2}{2K_2} t^{-n_2} h_2^2 + \frac{\alpha_1^2}{K_1} t^{-n_1} h_1 h_2\right) + \lambda^2 e^{rt} \frac{\alpha_1^2}{K_1} R_0 t^{m-n_1} h_1 + mR_0 t^{m-1}$.

Proof See Appendix "Derivation of Corollary 1".

We put forth that if there should be indeed a 'competitive or oligopolistic reaction' between the two competing firms, each firm will focus on how much innovation effort it devotes compared with its competitor's. Therefore, the relative innovation effort can be more important than an absolute level.

Theorem 2 (Oligopolistic reaction) *Firms' innovation strategy based on an oligopolistic reaction is not directly affected by their innovation knowledge levels, i.e.,* x_i*'s. More specifically, the relative innovation effort (defined as* u_1^*/u_2^**) is directly determined by each firm's innovation cost structure (i.e.,* C_1 *and* C_2*) and* $h_i(t)$*'s, both of which are influenced by firms' learning capabilities as captured in* n_1 *and* n_2.

Proof From (14) and (15), it is straightforward to show $\tilde{u} \equiv \frac{u_1^*}{u_2^*} = \frac{\alpha_1 h_1 K_2}{\alpha_2 h_2 K_1} t^{n_2 - n_1} = \frac{\alpha_1 h_1 C_2(t)}{\alpha_2 h_2 C_1(t)}$, from which the theorem can be derived.

Now we can return to the implication of h_i. From (19), $\frac{\partial J^i}{\partial x_i} = -\lambda e^{-\lambda(x_1 + x_2)} h_i = \lambda(1 - F_1)(1 - F_2)[-h_i]$ and thus, $\frac{\partial J^1/\partial x_1}{\partial J^2/\partial x_2} = \frac{-h_1}{-h_2}$. As discussed already, $-h_i$ can be regarded as a relative measure of the effectiveness of Firm i's technological knowledge in improving the firm's expected profit, compared with the competitor's.

3.1.3 Dynamic Causal Relation Analysis

The optimal control model developed in the previous section enables us to construct a dynamic causal relation map among key factors in the firms' innovation game. Figure 2 shows the causal chain between critical elements of the mathematical model.

The market reward for the innovation influences the determination of $h_i(t)$s, which in turn affect each other as in Theorem 1 and Corollary 1: $h_i(t)$ is an indirect (comparative) measure of the effectiveness of the firm's innovation knowledge. Moreover, since both $h_i(t)$s are influenced by the market reward structure (via $E_1(t)$ in Theorem 1 and $G_1(t)$ in Corollary 1), the rate at which the market reward shrinks, i.e., ϕ_m, impacts both firms' innovation strategies.

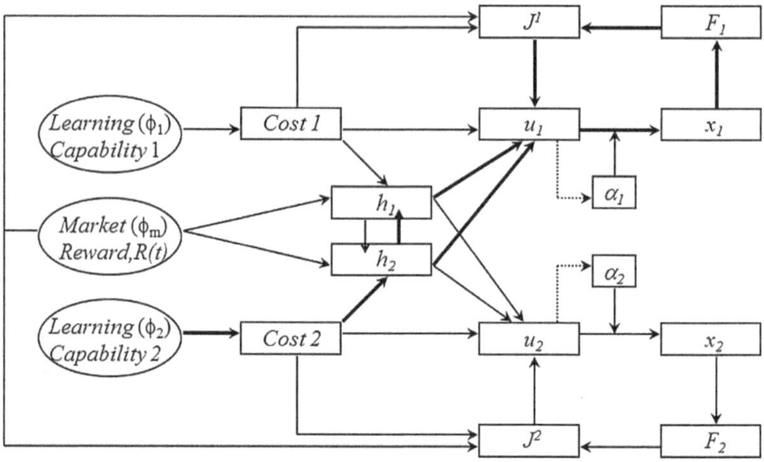

Fig. 2 Dynamic causal relationship

Each firm's learning capability directly affects its own innovation cost, i.e., ϕ_i directly determines C_i over time. Firm i's innovation strategy (u_i) is determined by its own cost (C_i), the marginal effectiveness of its own innovation knowledge (h_i), and also the marginal effectiveness of Firm j's innovation knowledge $\left(h_j\right)$. Therefore, Firm j's learning capability (via h_j) as well as the market reward (via h_i and h_j) plays an important role in Firm i's innovation decision-making.

Firm i's innovation strategy, i.e., u_i, the innovation effort level, determines its innovation knowledge stock at t, $x_i(t)$. As the knowledge stock increases, the chance of a successful innovation also grows, thus affecting F_i. The firm's profit function, J^i, is primarily determined by the interplay among market reward ($R(t)$), its own innovation cost (C_i), its probability of completing innovation (F_i), and the other firm's probability of completing innovation $\left(F_j\right)$ as in (7).

Although there are some differences in terms of the extent of direct influence, all the decision factors influence and are influenced by each other in complex causal ways. Figure 2 clarifies the process through which the decision factors evolve to interact with each other.

3.2 Numerical Examples

In order to see the dynamic interactions between firms' learning capabilities and market reward structure, we show numerical examples using the mathematical model set up previously. Table 2 shows the base case of parameter values for the numerical examples. The innovation game is supposed to span for 10 time units, i.e., $[1, 11]$: as discussed in association with (5), we are assuming that after $t = 11$,

Table 2 Base parameter values for numerical examples

α_1	α_2	K_1	K_2	ϕ_1	ϕ_2	ϕ_m	λ	r	R_0	T
1	*0.9*	*1*	*1.5*	*1.0*	*0.9*	*1.0*	*0.01*	*0.05*	*30*	*11*

the innovation will be outdated and have little value. In addition, we make the following assumptions.

- Initially compared with Firm 2, Firm 1 is more effective in converting innovation effort into technological knowledge ($\alpha_1 > \alpha_2$), and enjoys a more advantageous innovation cost structure ($K_1 < K_2$). Innovation costs will be discounted according to the market interest rate, $r = 0.05$.
- But, Firm 1 does not have any potential to become more efficient through learning over time (i.e., $\phi_1 = 1.0$). In order to examine the effect of two firms' learning rate difference on their innovation strategy, we will vary Firm 2's learning rate, ϕ_2.
- We will investigate the impact of market reward depreciation rate on firms' innovation strategy by varying ϕ_m, with the base value of $\phi_m = 1.0$, that is, no depreciation assumed initially.
- The innovation success probability is assumed to have a parameter $\lambda = 0.01$, which roughly implies that the completion of the innovation requires 100 ($= 1/\lambda$) units of technological knowledge on average.
- Finally, the lump sum value of market reward for the innovation is $R_0 = 30$. That is, if either firm succeeds in innovating at $t = 1$, i.e., at the very beginning of the innovation game, then the winning firm will earn 30 from the market. The market reward will not be discounted according to the market interest rate. Rather, it will be depreciated, abiding by ϕ_m.

We summarize the numerical examples in Figs. 3, 4 and 5. Figure 3 illustrates the impact of Firm 2's learning rate on Firm 1's innovation effort in terms of a comparative unit, i.e., u_1/u_2. Figure 4 shows the impact of market reward depreciation rate on Firm 1's innovation strategy. Finally, in Fig. 5, we construct the cumulative probability function for completing the innovation as the market depreciation rate changes.

From the numerical examples, we can make critical inferences.

- First, as Fig. 3 indicates, a firm's learning capability not only determines its own innovation cost structure, but also significantly affects its competitor's innovation strategy. When both firms' learning capabilities were similar (i.e., $\phi_2 = 1.0$), the ratio of both firms' innovation efforts (i.e., u_1/u_2) stayed almost constant or increased very slightly over time, determined by the initial cost structures: initially, Firm 1 enjoyed a lower innovation cost, and thus, it seemed to devote more resources to the innovation than the other firm, Firm 2, did. As Firm 2 became more efficient due to its higher learning capability, Firm 1's innovation effort decreased significantly: for $\phi_2 = 0.8$, despite its initial cost advantage over Firm 2's, Firm 1 started spending less from around $t = 5$, that is,

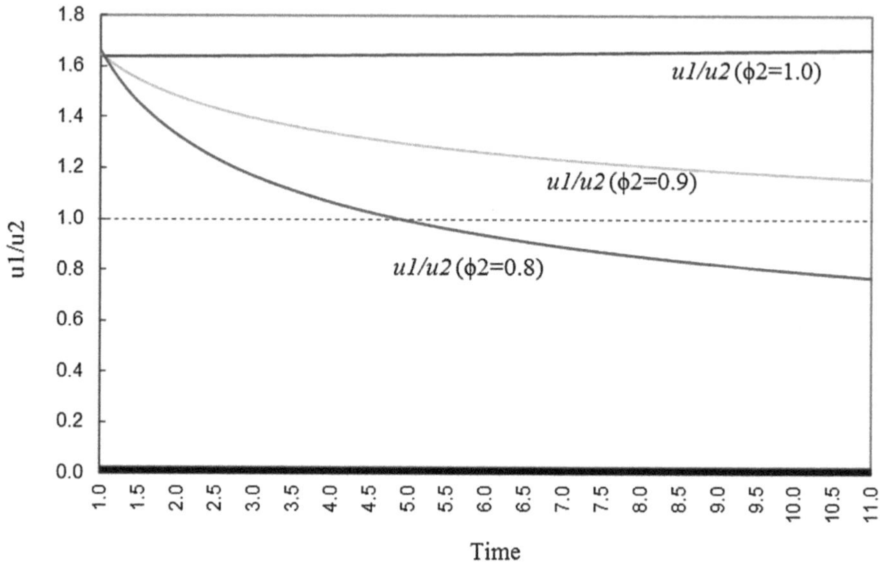

Fig. 3 Impact of learning rate changes on competitive innovation strategy

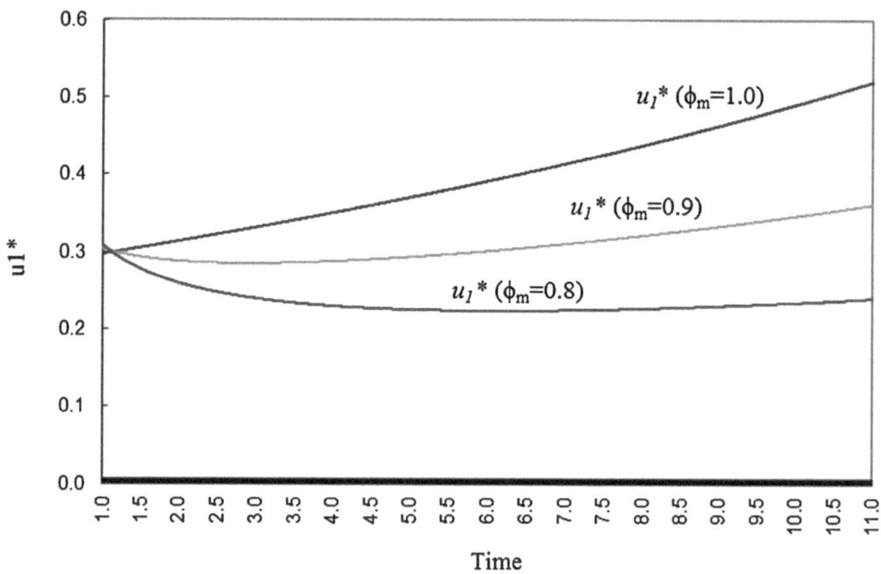

Fig. 4 Impact of market reward structure on firm's innovation strategy

$u_1/u_2 < 1.0$ from $t = 5$. The result shown in Fig. 3 was already anticipated in Theorem 2, where it was proved that a firm's oligopolistic innovation strategy is influenced by the innovation cost structures as well as learning capabilities.

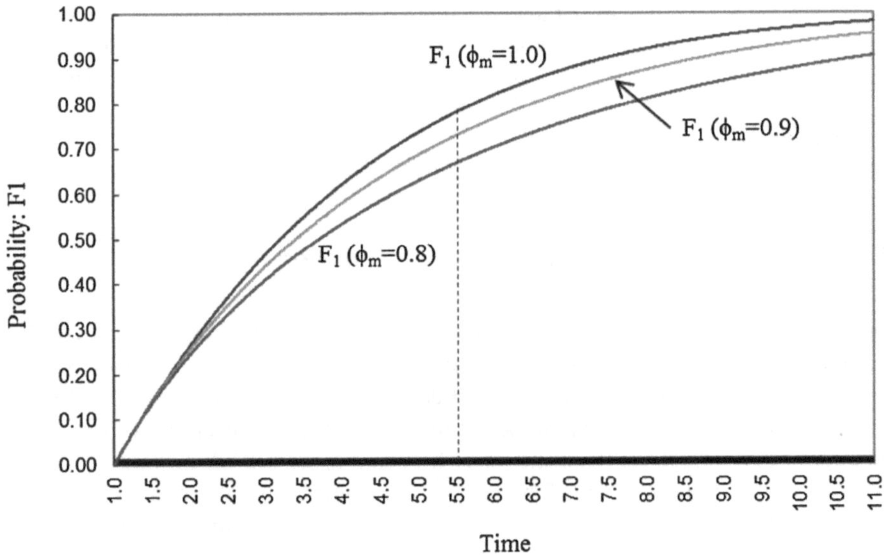

Fig. 5 Impact of market reward structure on innovation success probability

- The depreciation rate of market reward, ϕ_m, seemed to affect the firm's innovation strategy significantly. In Fig. 4, we can see that Firm 1 increased its innovation effort as time advanced when the depreciation rate was not so high, e.g., $\phi_m = 1.0$ or 0.9, although it devoted more resources to the innovation when the depreciation rate was lower, e.g., $\phi_m = 1.0$, than when it was relatively high, e.g., $\phi_m = 0.9$. When the market sharply reduced its reward as the innovation completion time delayed, e.g., $\phi_m = 0.8$, the firm actually decreased its innovation effort throughout the decision time horizon.
- Additional numerical tests showed that changes in a firm's learning capability did not have a bigger impact on the other firm's innovation success probability than did the market reward depreciation rate. For instance, the effect of Firm 2's learning rate change on Firm 1's innovation success probability was not as much as that of market reward depreciation rate on Firm 1's success probability. From the dynamic causal map in Fig. 2, we can infer that the market reward might have more ways to influence firms' stocks of innovation knowledge than each firm's learning capability might have. The learning capabilities impact the innovation knowledge via longer and more indirect links. Figure 5 presents the changes of Firm 1's innovation success probability as the market reward depreciation rate changes: at about half the decision horizon, i.e., $t = 5.5$, 20% difference in the depreciation rate (e.g., $\phi_m = 1.0$ versus $\phi_m = 0.8$) made Firm 1's innovation success probability differ by more than 10%, which should be consequential in the competitive innovation game.

We tried a wide range of parameter values, and derived similar dynamic patterns as observed in the figures. The importance of such numerical examples is to confirm our proposition that firms' learning capabilities and the market reward structure should be considered concurrently in order to understand the innovation competition more robustly.

3.3 *Managerial Implications and Discussion*

We have investigated a dynamic interplay between competing firms' learning capabilities, market reward structure for the innovation, and other cost factors in an innovation game setting. As alluded earlier, we could find many real-world situations where our analysis might be applicable at least conceptually. One such example is the pharmaceutical industry. Innovation for new drug development is very important in the industry, forcing drug companies to spend millions of dollars each year in R&D. Since developing a new drug requires a huge investment, i.e., it is an extraordinarily research-intensive process, there are only a few very resourceful drug companies for each innovation competition. A drug company that competes in the innovation race will be able to recover its investment only when it becomes the first to successfully complete the project for the new drug. That is, competition in the pharmaceutical industry can be characterized by the 'winner-takes-all' reward system. This particular reward structure is due to the industry's patent system that recognizes only the first inventor. Moreover, the significant R&D intensity along with the learning-by-doing effect in the process makes the firm's learning capability critical in successfully winning the innovation game.

In this section, we have established a differential games model, developed a dynamic causal map, and presented numerical examples so as to clarify the dynamic causal relationship among key decision elements. Previous studies on the similar subject mainly ignored the firms' long-term organizational competence such as learning capability when analyzing game-theoretic innovation races. That is, the firms' innovation potentials were regarded as fixed or static over time. By allowing differentiated learning capabilities in the analysis, we were able to understand a capability-based competition between firms in the innovation game, and extended the research perspective in this area. Another difference of our analysis from the previous ones is that we took into account a dynamic nature of the market reward for the innovation: whereas previous studies considered only a fixed amount of reward for the innovation, we made a more realistic assumption that the market reward decreases as the firms' completion of the innovation delays. Our analysis result can provide managers with dynamic insights when designing an innovation strategy, by helping them understand the necessity of taking into consideration their competitors' organizational competence in addition to their own current cost structure.

4 Firms' Cooperation and Competition for Innovation[2]

Now we investigate under what circumstances firms effectively participate in creating an industry-wide infrastructure such as an industry standard without incurring opportunistic behaviors. Literature attests that free riding is unavoidable when companies are involved in offering public goods, which will eventually benefit all of the members in the industry regardless of whether they actually contributed to making the goods. That is, the steady-state equilibrium stocks of capital investment for a Markov perfect Nash equilibrium are lower than those in the collusive case, where the companies act like a single entity, not competitors. We postulate that in addition to the non-discriminating 'industry-wide, i.e., common' benefits, the firm participating in the process can earn another more important type of benefits, firm-specific benefits, which improve the firm's own internal operations and enhance its performance. Our analysis shows that if it is indeed possible for the firms to realize there exist both common and firm-specific benefits in collaboration and to be able to capitalize on both benefits simultaneously, the opportunistic behaviors can be mitigated or even eliminated, meaning economically better solutions can be derived from both the individual firm's perspective and the entire industry's.

4.1 Competition Versus Collaboration

Collaboration between firms seems prevalent these days: even collaboration between competitors in the same industry is not unusual (Hamel et al. 1989, Park et al. 2014). Why do firms collaborate? There are many reasons for firms to collaborate: sometimes they collaborate to build an infrastructure in the industry, e.g., a common industry standard, to jointly develop a new product or technology, to conduct a research project together, so forth (Cellini and Lambertini 2009; Youssef et al. 2013). The outputs of such collaborations, however, have one thing in common: they have the characteristics of 'public goods.' For example, the benefits from an industry-wide infrastructure are nondiscriminatory, i.e., everyone in the industry might get the benefits regardless of whether they have actually participated in such an effort.

When the investment is involved with such characteristics as public goods, there is usually an issue of free-rider or opportunism, i.e., some companies enjoy the infrastructure's benefits without paying appropriate dues. Economically speaking, the total investment level is lower than that under an optimal condition with no free-rider problem, i.e., a collusive case.

[2]This section is a refined version of Kim B. and S. Lee (2005) "Firms' collaboration on building a business infrastructure as public goods: dynamics of common versus firm-specific benefits." *Optimal Control Applications and Methods, 26* (5), 265–290.

But, are the public or common benefits the only gains a participating firm earns from such collaboration? We put forth that while collaborating with other companies on building an industry-level infrastructure, the company will learn technical knowledge as well as know-how that can be employed to improve its own internal manufacturing process and enhance its own productivity in-house without much extra cost. That is, the firm gets not only the industry-level ‘public or common’ benefits, but also its own ‘firm-specific’ benefits from the collaboration. In this section, we look into whether the opportunistic behavior becomes more or less prevalent in the new setting by comparing the steady-state total investment levels.

4.2 A Differential Game Model

In our differential game model, we include ‘firm-specific’ benefits as well as ‘public goods,’ i.e., common benefits, in the process of creating a common business infrastructure by firms potentially competing in the industry. For analytical tractability, we assume a differential game setting, where two firms are collaborating on building a common infrastructure.

4.2.1 A Markov Perfect Nash Equilibrium Case

Suppose there are two players, i.e., firms, in the industry, $i = 1, 2$. We put forth a ‘linear-quadratic’ model to depict the collaboration game:

$$\begin{aligned} &\mathit{Maximize}\, J^i = \int_0^\infty e^{-rt}\left\{(x_1 + x_2)[a_i - (x_1 + x_2)] + s_i x_i[b_i - x_i] - \rho u_i - \frac{1}{2}u_i^2\right\}dt \\ &\mathit{Subject\ to}\, \dot{x}_1 = u_1 - \delta x_1 \\ &\qquad\qquad\quad \dot{x}_2 = u_2 - \delta x_2. \end{aligned}$$

The state variable, $x_i(t)$, stands for firm i’s cumulative investment in the common infrastructure by time t. It accumulates following a dynamic equation, $\dot{x}_i = u_i - \delta x_i$, where $u_i(t)$ is the firm’s current investment at t and $\delta \geq 0$ a constant (a decaying rate). Therefore, $x_1 + x_2$ is the total investment accumulated until t by the two firms for the common infrastructure. Each firm gains $(x_1 + x_2)[a_i - (x_1 + x_2)]$ from the infrastructure. The firm-specific benefit from the infrastructure is captured by $s_i x_i[b_i - x_i]$, where s_i is 1 if there is the firm-specific benefit and 0 if not. Investing $u_i(t)$ in building the infrastructure costs the firm $\rho u_i + \frac{1}{2}u_i^2$ at t. In addition, $r \geq 0$ is the discount rate, $a_i > 0$, and $b_i > 0$. Since we focus on the case where both firm-specific and common benefits are present, we use $s_1 = s_2 = 1$ for the ensuing formulation. See Example 1 for a practical case.

Example 1 Practical applications of common versus company-specific benefits

We want to establish a relationship between the analysis model and the actual case in the industry. Suppose a company has to make a decision on whether it participates in a certain industry-wide collaboration. In order to make such a decision, the company has to understand what costs as well as benefits are involved. It is relatively straightforward for the company to figure out the industry-wide benefits represented by $(x_1 + x_2)[a_i - (x_1 + x_2)]$ in the game-theoretic model. For instance, these benefits may be derived from the overall efficiency increase in the industry due to, say, standardization of components or parts, which enables the company and its suppliers to communicate with each other far more effectively. However, the company is also aware that these benefits are common, i.e., enjoyed by other companies in the industry regardless of whether they actually contributed to the collaboration. These are the potential costs of such collaboration. Further suppose that given these industry-wide benefits and the associated costs due to free riding, the company still cannot make a definitive decision. Then, the company realizes that there are additional benefits embodied in $x_i[b_i - x_i]$, which are firm-specific since only the company can reap these benefits by applying what it learns from the collaboration to improving its internal operations. For instance, as it tries to implement the industry standardization, the company learns how to organize complicated code numbers and utilizes the newly gained knowledge to improve its own inventory management. Now should it participate in the industry-wide collaboration? Our analysis endeavors to help top managers at the company make an optimal decision on collaboration.

For $i = 1$ and the indicated infinite time case, the Hamilton–Jacobi–Bellman (HJB) equation becomes:

$$
\begin{aligned}
rJ = \max_{u_1}\Big\{ & (x_1 + x_2)[a_i - (x_1 + x_2)] + x_i[b_i - x_i] - \rho u_i - \frac{1}{2}u_i^2 + J_{x_1}(u_1 - \delta x_1) \\
& + J_{x_2}(u_2 - \delta x_2)\Big\}
\end{aligned}
\tag{20}
$$

To achieve the indicated maximization, we set the partial derivative of (20) with respect to u_1 equal to zero and thus obtain $u_1 = J_{x_1} - \rho$. Similarly, we can get $u_2 = J_{x_2}^2 - \rho$ for $i = 2$. In order to solve the dynamic HJB equation, we use the following value functions:

$$J = \frac{\alpha}{2}(x_1 + x_2)^2 + \beta_1(x_1 + x_2) + \frac{\theta}{2}x_1^2 + \phi_1 x_1 + \gamma_1 \text{ for, } i = 1, \text{ and}$$

$$J^2 = \frac{\alpha}{2}(x_1 + x_2)^2 + \beta_2(x_1 + x_2) + \frac{\theta}{2}x_2^2 + \phi_2 x_2 + \gamma_2 \text{ for } i = 2,$$

where $\alpha, \beta_i, \theta, \phi_i$, and γ_i are constants to be determined.

Theorem 3 *Using the value function suggested above and the standard dynamic optimization method, we obtain a solution to the Markov perfect Nash equilibrium case as follows:*

$$x_1 = c_1 e^{(2\alpha+\theta-\delta)t} + c_2 e^{(\theta-\delta)t} + \frac{\alpha(\beta_2+\phi_2-\rho) - (\alpha+\theta-\delta)(\beta_1+\phi_1-\rho)}{(\alpha+\theta-\delta)^2-\alpha^2}$$
$$x_2 = c_1 e^{(2\alpha+\theta-\delta)t} - c_2 e^{(\theta-\delta)t} + \frac{\alpha(\beta_1+\phi_1-\rho) - (\alpha+\theta-\delta)(\beta_2+\phi_2-\rho)}{(\alpha+\theta-\delta)^2-\alpha^2}. \tag{21}$$

For a steady-state solution, we require that $2\alpha+\theta-\delta<0$ *and* $\theta-\delta<0$, *where*

$$\theta = \frac{(r+2\delta) - \sqrt{(r+2\delta)^2+8}}{2},\ \alpha = \frac{(r+2\delta-2\theta) - \sqrt{(r+2\delta-2\theta)^2+24}}{6},$$
$$\phi_1 = \frac{b_1-\theta\rho}{r+\delta-\theta},\ \phi_2 = \frac{b_2-\theta\rho}{r+\delta-\theta},\ W_i = \frac{\alpha(2r\rho+2\delta\rho-b_1-b_2)}{r+\delta-\theta} - a_i,$$
$$\beta_1 = \frac{2\alpha+\theta-r-\delta}{(2\alpha+\theta-r-\delta)^2-\alpha^2}W_1 + \frac{-\alpha}{(2\alpha+\theta-r-\delta)^2-\alpha^2}W_2,$$
$$\beta_2 = \frac{-\alpha}{(2\alpha+\theta-r-\delta)^2-\alpha^2}W_1 + \frac{2\alpha+\theta-r-\delta}{(2\alpha+\theta-r-\delta)^2-\alpha^2}W_2. \tag{22}$$

Once initial values of the state variables ($x_1(0) = x_{10}$ *and* $x_2(0) = x_{20}$) *are given, we can determine* c_1 *and* c_2. *Since we are concerned with the steady-state 'investment level,' however, we do not intend to specify such values.*

Proof See Appendix "Derivation of Theorem 3".

For future reference, we define $\lim_{t\to\infty} x_i(t) \equiv x_{is}^N$ as the firm i's steady-state cumulative investment. Therefore,

$$\lim_{t\to\infty} x_1(t) \equiv x_{1s}^N = \frac{\alpha(\beta_2+\phi_2-\rho) - (\alpha+\theta-\delta)(\beta_1+\phi_1-\rho)}{(\alpha+\theta-\delta)^2-\alpha^2}, \text{ and}$$
$$\lim_{t\to\infty} x_2(t) \equiv x_{2s}^N = \frac{\alpha(\beta_1+\phi_1-\rho) - (\alpha+\theta-\delta)(\beta_2+\phi_2-\rho)}{(\alpha+\theta-\delta)^2-\alpha^2},$$

where N stands for 'Nash equilibrium' and s for 'steady state.'

4.2.2 A Perfect Coordination (Collusive) Case

Now consider a perfect coordination case, where the two firms make a decision in harmony as if they were under a single decision-making authority, but the two firms are still separate and independent entities.

The optimal control theory problem for this situation is formulated as follows:

$$\textit{Maximize}\, J = \int_0^\infty e^{-rt}\{(x_1+x_2)[a_1+a_2-2(x_1+x_2)]+x_1[b_1-x_1]+x_2[b_2-x_2]$$
$$-\rho(u_1+u_2)-\frac{1}{2}\left(u_1^2+u_2^2\right)\}dt$$

$$\textit{Subject to}\ \dot{x}_1 = u_1-\delta x_1$$
$$\dot{x}_2 = u_2-\delta x_2.$$

Applying the optimal control theory, we use the current value Hamiltonian to solve the problem.

$$H = (x_1+x_2)[a_1+a_2-2(x_1+x_2)]+x_1[b_1-x_1]+x_2[b_2-x_2]-\rho(u_1+u_2)$$
$$-\frac{1}{2}\left(u_1^2+u_2^2\right)+\lambda_1(u_1-\delta x_1)+\lambda_2(u_2-\delta x_2) \tag{23}$$

Necessary conditions are: $\frac{\partial H}{\partial u_1} = -\rho-u_1+\lambda_1 = 0,\ u_1 = \lambda_1-\rho;$ $\frac{\partial H}{\partial u_2} = -\rho-u_2+\lambda_2 = 0,$ $u_2 = \lambda_2-\rho;$
$\dot{\lambda}_1 = r\lambda_1-\frac{\partial H}{\partial x_1} = (r+\delta)\lambda_1+6x_1+4x_2-(a_1+a_2+b_1).$

Likewise, $\dot{\lambda}_2 = r\lambda_2-\frac{\partial H}{\partial x_2} = (r+\delta)\lambda_2+4x_1+6x_2-(a_1+a_2+b_2).$

Theorem 4 *The steady-state solution of the state variables for the perfect coordination case, i.e.,* x_1 *and* x_2*, is as follows:*

$$\lim_{t\to\infty} x_1(t) \equiv x_{1s}^C = \frac{4[\rho(r+\delta)-(a_1+a_2+b_2)]-[\rho(r+\delta)-(a_1+a_2+b_1)][\delta(r+\delta)+6]}{[\delta(r+\delta)+6]^2-16},\ \text{and}$$
$$\lim_{t\to\infty} x_2(t) \equiv x_{2s}^C = \frac{4[\rho(r+\delta)-(a_1+a_2+b_1)]-[\rho(r+\delta)-(a_1+a_2+b_2)][\delta(r+\delta)+6]}{[\delta(r+\delta)+6]^2-16}. \tag{24}$$

As in the Nash equilibrium case, s stands for 'steady state' and C for 'coordination.'

Proof See Appendix "Derivation of Theorem 4".

Corollary 2 *Firm 1's steady-state cumulative investment under the collusive decision-making scheme, i.e.,* $\lim_{t\to\infty} x_1(t) \equiv x_{1s}^C$*, is larger than Firm 2's if and only if*

$(b_1 - b_2)[\delta(r+\delta)+10] > 0$, i.e., $b_1 > b_2$. That is, $x_{1s}^C > x_{2s}^C$ iff $b_1 > b_2$. It implies that the steady-state investment level is determined by the magnitudes of b_i.

Here we need to understand the meaning of b_i. As alluded already, Firm i gets $x_i(b_i - x_i)$ as firm-specific benefits, which reaches its maximum of $\frac{b_i^2}{4}$ at $x_i = \frac{b_i}{2}$. It is also implicitly assumed that $0 \le x_i \le b_i$, that is, the maximum of accumulated investment by Firm i is b_i: it is possible that $x_i > b_i$, but then $x_i(b_i - x_i) < 0$ and therefore there is an incentive for the firm to keep its cumulative investment below b_i, other things being equal. The bigger b_i, the more potential the firm has to earn 'firm-specific' benefits. We can draw similar implications for a_i.

It is also interesting to note that if there are no firm-specific benefits, i.e., $s_1 = s_2 = 0$, then the two firms will invest the same amount in the long run, i.e., $x_{1s}^C = x_{2s}^C$: see Appendix 3. It is interesting because it holds regardless of the magnitudes of a_i.

4.3 Analysis of the Model

In this section, we more thoroughly analyze the theorems, to explore our main question about free riding in collaboration. Dockner et al. (2000) showed the solutions for the differential game problem with public benefits only and at the total investment level, not differentiated between two firms. Table 6 in Appendix 3 summarizes those results. Our approach to solving the problem is different from that in Dockner et al. (2000) twofold. First, we have calculated the individual firm's investment level, not just the summed one. Thus, it becomes possible for us to see how the difference between the firms' investment levels is caused as shown in the corollary. Second, we have considered two cases separately, i.e., the case with common benefits only and that with firm-specific benefits in addition to the common ones. Table 7 in Appendix 3 summarizes our results.

In Table 3, we define additional notations to compare the steady-state investment levels in different situations. X_{1N} is the steady-state total investment (including both firms' investments) in building the infrastructure, when there exist common benefits only and the two firms are competing against each other (i.e., Nash equilibrium). X_{1C} is the same as X_{1N} except that the two firms coordinate their decision-making, i.e., a collusive situation. For X_{1N} and X_{1C}, '1' signifies that there

Table 3 Steady-state total investment levels

	A differential game situation (A Markov perfect Nash equilibrium)	A collusive situation (A perfect coordination)
Common (public) benefits only	X_{1N}	X_{1C}
Common and firm-specific benefits	X_{2N}	X_{2C}

is only 'one' kind of benefit, i.e., the common benefits, from participating in the infrastructure building.

When there are two different types of benefits (i.e., firm-specific and common) coexist, we use X_{2N} and X_{2C} to represent the steady-state total investment summing the two firms' together: X_{2N} is for the Nash equilibrium, whereas X_{2C} for the collusive case.

Using Table 7 in Appendix 3, we derive the exact values of X_{1N} and X_{1C}:

$$X_{1N} = \frac{2\rho(r+\delta-\alpha)-(a_1+a_2)}{(r+\delta-3\alpha)(2\alpha-\delta)} \text{ and } X_{1C} = \frac{2[(a_1+a_2)-\rho(r+\delta)]}{\delta(r+\delta)+8}.$$

Employing the notation in the previous section (see (21) and (24)), we rewrite X_{2N} and X_{2C} as follows:

$$X_{2N} \equiv x_{1s}^N + x_{2s}^N = \frac{(a_1+a_2)(r+\delta-\theta)-(2r\rho+2\delta\rho-b_1-b_2)(r+\delta-\alpha-\theta)}{(r+\delta-\theta)(3\alpha+\theta-r-\delta)(2\alpha+\theta-\delta)},$$

$$X_{2C} \equiv x_{1s}^C + x_{2s}^C = \frac{2[(a_1+a_2)-\rho(\gamma+\delta)]+(b_1+b_2)}{\delta(r+\delta)+10}.$$

Observation 1 $X_{1N} \leq X_{1C}$. It implies that the coordinated investment decision brings about a steady-state investment level higher than when the decision is made under a Markov perfect Nash equilibrium, i.e., when competing firms invest in public goods, the free-rider problem is unavoidable.

Proof See Appendix "Derivation of Observation 1".

A key managerial implication of Observation 1 is that if there are only the common 'industry-wide' benefits in collaboration, opportunistic behaviors among the participating firms are unavoidable. That is, the collaboration is ineffective from the entire industry's (or economy's) perspective. This ineffectiveness is in turn conducive to the individual firm's reluctance to participate in such collaboration.

Observation 2 A necessary condition for $X_{1C} \leq X_{2N}$. It is possible that $X_{1C} \leq X_{2N}$, if it holds $X_{2N} - X_{1C} = [\delta(r+\delta)+8]\cdot[(a_1+a_2)(r+\delta-\theta)-(2\rho(r+\delta)-b_1-b_2)(r+\delta-\alpha-\theta)] - 2[(a_1+a_2)-\rho(r+\delta)]\cdot(r+\delta-\theta)(3\alpha+\theta-r-\delta)(2\alpha+\theta-\delta)\geq 0$, which can be rearranged as $(b_1+b_2) = C(a_1+a_2)+D$, where $C = \frac{(r+\delta-\theta)\{2(3\alpha+\theta-r-\delta)(2\alpha+\theta-\delta)-[\delta(r+\delta)+8]\}}{(r+\delta-\alpha-\theta)[\delta(r+\delta)+8]}$ and

$$D = \frac{2\rho(r+\delta)\{(r+\delta-\alpha-\theta)[\delta(r+\delta)+8]-(r+\delta-\theta)(3\alpha+\theta-r-\delta)(2\alpha+\theta-\delta)\}}{(r+\delta-\alpha-\theta)[\delta(r+\delta)+8]}.$$

This is the quintessential outcome of our analysis. It shows that under certain circumstances that largely depend on the relative magnitudes of a_i and b_i, the

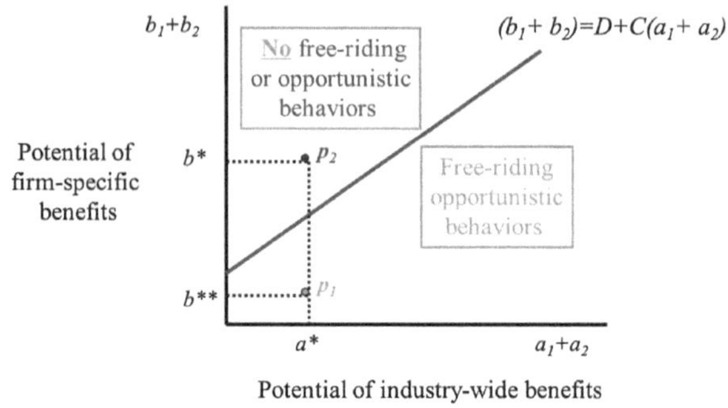

Fig. 6 Managerial implications of Observation 2

steady-state total investment of a Markov perfect Nash equilibrium can be higher than that in a perfect coordination case.

Observation 2 attests that if there exist both the industry-wide and the firm-specific benefits simultaneously, collaboration between competing firms can be more effective than when the firm-specific benefits are absent. That is, the opportunistic behaviors among the participating firms can be lessened or even eliminated. As shown in Observation 1, this is not possible if the firm-specific benefits do not exist.

We can draw the condition in Observation 2 as in Fig. 6. If the *combined* potential of firm-specific benefits (i.e., b_1+b_2) is sufficiently large compared with that of industry-wide benefits (i.e., a_1+a_2), it is possible to circumvent the free-riding or opportunistic behaviors. Consider two points p_1 at (a^*, b^{**}) and p_2 at (a^*, b^*), where $b^* > b^{**}$: the free riding cannot be avoided at p_1, whereas it can be avoided at p_2.

Observation 3 A necessary condition for $[X_{1C} \leq X_{2C}]$ It holds if $b_1+b_2 \geq \frac{4[(a_1+a_2)-\rho(r+\delta)]}{\delta(r+\delta)+8}$, which is equivalent to $\frac{b_1+b_2}{2} \geq X_{1C}$.

Observation 3 means that if the combined potential of firm-specific benefits is large enough, more specifically larger than twice the steady-state total investment of the collusive case with the industry benefits only (i.e., $b_1+b_2 \geq 2X_{1C}$), it holds that $X_{1C} \leq X_{2C}$, implying that the steady-state total investment of the collusive case with both types of benefits is larger than that with the industry-wide benefits only. On the contrary, one can say that unless the firm-specific benefits are large enough, the collusive decision-making under the industry-wide benefits only can be better or more desirable.

4.4 Numerical Examples and Inferences

In order to more clearly see the implications of our analysis, we conduct numerical analysis using the parameter values in Table 4.

Table 4 Base values of parameters in the numerical analysis

r	δ	ρ	s_1	s_2	a_1	a_2	b_1	b_2
0.1	0.2	1	1	1	1.1	1	0.5	0.5

In Fig. 7, note that Observation 2 is valid only when $b = b_1 + b_2$ is not so small relative to $a = a_1 + a_2$. It indicates that only when the firm-specific benefits are significant in relation to the common benefits, it is possible to avoid the free-riding problem: it is consistent with the managerial implications inferred in Fig. 6. Figure 8 shows the cases where $X_{1N} \leq X_{2N}$ and $X_{2C} \leq X_{2N}$.

From the first and the second observation, we are able to suggest that under an appropriate condition, it is possible to have $X_{1N} \leq X_{1C} \leq X_{2N}$. Its key implication is that if there are firm-specific benefits when firms work together to build a business infrastructure for common benefits, the coordination can be more desirable from the whole economy's perspective: that is, the total investment in building the infrastructure can be greater than that for a complete coordination case, where there may be no competition between the firms. Therefore, the free-rider problem can be mitigated.

Figure 9 recapitulates the key observations of our model in an integrated way. Based on the numerical analysis results, we divide the $a - b$ space into 5 regions, R_1, R_2, R_3, R_4, and R_5. There is an extreme case, R_5, where the common benefit is very large, but the firm-specific benefit is very small. Thus, in R_5 it is difficult to avoid the free-riding problem, implying that the collusive case offers the largest steady-state investment in creating the public goods. Also in R_3 and R_4, the firm-specific benefit is not big enough for the competitive setting to avoid free riding: the main difference between R_3 and R_4, though, is that in R_3 the firm-specific benefit is larger than that in R_4 so that the existence of firm-specific benefits is better

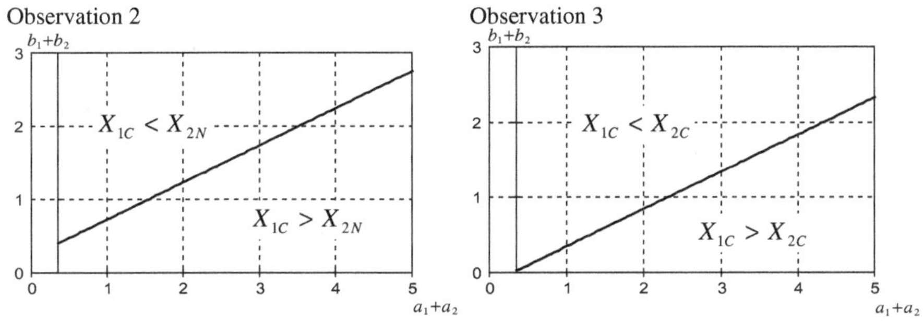

Fig. 7 Observations 2 and 3

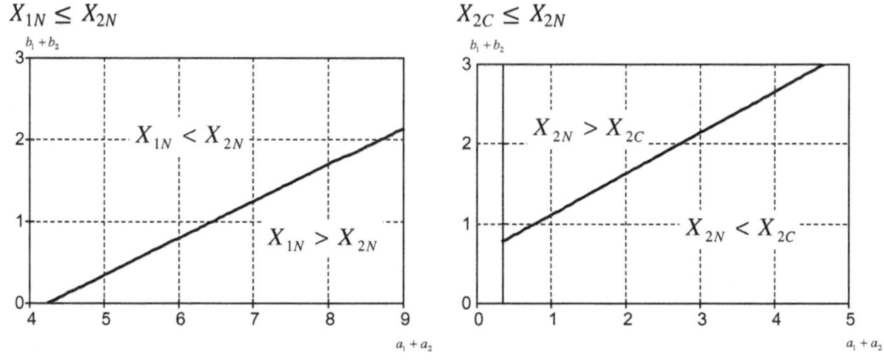

Fig. 8 Cases where $X_{1N} \leq X_{2N}$ and $X_{2C} \leq X_{2N}$

Fig. 9 Combined observations

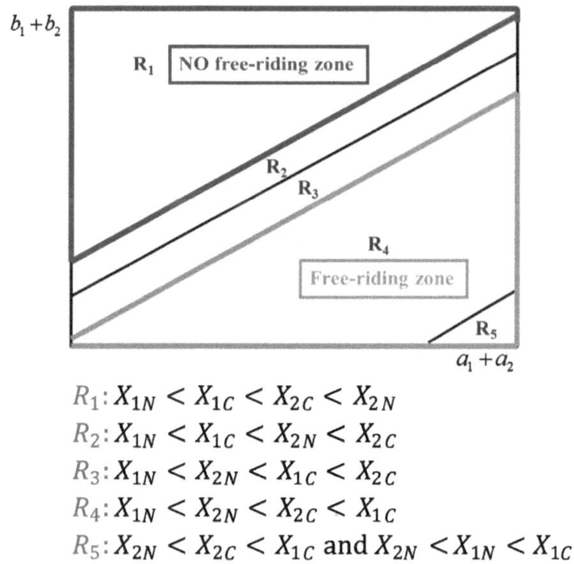

for the economy as a whole than when there are only the common benefits even under the collusive situation.

Finally, in R_1 and R_2, the firm-specific benefit becomes large enough to mitigate the free riding problem, i.e., the total investment level becomes larger than that of the collusive case. The important difference between R_1 and R_2 is that in R_1 the firm-specific benefit becomes sufficiently large so that the cumulative investment level of the competitive setting is larger than that in any other situation including both 'common benefit only' collusive case and 'common plus firm-specific' collusive case.

Figure 9 again underlines our primary observation: if the firm-specific benefit is sufficiently large compared with the common benefit, it will be possible to avoid the free riding problem even when there is no collusive arrangement between the firms. From an individual firm's perspective, it is also an important conclusion. If the company can figure out how to utilize knowledge derived from the 'common effort' to improve its own internal operations, it will be more willing to get involved in developing the industry-wide infrastructure.

4.5 Managerial Implications

We started with a question "Why do firms collaborate on building a common industry-level infrastructure?" It is also related with managerial issues such as coordination and alliances between (sometimes competing) firms. The industry-level infrastructure acts like public goods in that it offers benefits to most of the firms in the industry, regardless of whether the beneficiaries have contributed to building the infrastructure in the first place. Thus, an issue of 'free riding' is a serious one. Fershtman and Nitzan (1991) showed that when providing a public good, a collusive case generates more investment than a (Markov perfect) competitive case, implying that the free rider problem hampers achieving an optimal investment level.

From a managerial viewpoint, we put forth that the firms get not only 'common,' but also some 'firm-specific' benefits by participating in developing the public goods. For instance, by participating in creating an industry-wide IT infrastructure, the firm will be able to *not only* do business with other partners in the industry more efficiently, *but also* improve its internal manufacturing process so as to enhance its own productivity significantly. Should the firm understand this mechanism, it will be more willing to collaborate with other firms on providing public goods (such as common infrastructure or industry-wide standardization), which will benefit the economy as a whole in the end.

Including both common and firm-specific benefits simultaneously in the model, we were able to prove that under certain circumstances, it is indeed possible to avoid the free riding problem: such circumstances, in large part, depend on the balance of difference between sizes of common and firm-specific benefits: if the firm-specific benefit is sufficiently large compared with the common benefit, it will be possible to avoid the free riding problem even when there is no collusive arrangement between the firms. It has a profound implication to an individual firm. If the company is capable of finding out how to utilize knowledge derived from the 'common effort' so as to enhance its own internal productivity, it can afford to get *more* involved in developing the common infrastructure than otherwise. It will have a positive repercussion in the economy as a whole.

Following the same reasoning in reverse, one can state that unless there are firm-specific benefits in the inter-firm collaboration, any such endeavor is unsustainable, implying that it can be affected by some form of opportunism, i.e., 'free riding.' That is, only when external collaboration also enables the firm to enhance its internal learning as 'firm-specific' benefits, it becomes meaningful for the firm to participate in such collaboration.

Exercise Problems

1. There are two firms that invest in the common infrastructure. Let $u_i(t)$ be the investment of Firm i in the common infrastructure at time t and $x(t)$ the cumulative investment at time t, which evolves according to $\dot{x}(t) = u_1(t) + u_2(t)$. Investment cost of Firm i is given by $a_i(u_i(t))^2/2$. If each firm's benefit increases proportionally to the cumulative investment and the discounting effect can be disregarded, the objective functions are given by $J^i = \int_0^T \left[x(t) - \frac{a_i(u_i(t))^2}{2}\right] \mathrm{d}t$. Show that the open-loop Nash equilibrium is the same as the feedback Nash equilibrium.
2. Two firms compete for the development of a breakthrough technology. Different from the 'winner-takes-all' situation, assume that the success of technology development also benefits the competing firm to some extent. Specifically, when one firm succeeds in the technology development, the other firm stops technology investment and sells a "me-too" product by copying the developed technology. Denote the present value of the winner's profit at the completion time t by P_w and the present value of the loser's profit at t by P_l. Note that P_w is larger than P_l due to the first-mover advantage. Firm i determines its innovation effort $u_i(t)$, which costs $(u_i(t))^2/2$, but contributes to the accumulation of Firm i's technological knowledge for innovation $x_i(t)$, i.e., $\dot{x}_i(t) = u_i(t)$. The probability of successful development of the breakthrough technology by time t is given by $F_i(x_i(t)) = 1 - e^{-\lambda x_i(t)}$. A discounting rate is r and $x_i(0) = x_0$. Construct the model for this innovation game for the planning period $[0, T]$.
3. The cost term in (7) is represented as $e^{-rt}(1 - F_1(x_1))(1 - F_2(x_2))C_1(t)\frac{u_1^2}{2}$. In other words, Firm 1's innovation cost at t occurs only when neither firm has succeeded in the innovation by t. Do you think that this assumption makes sense? Discuss market conditions that make this assumption valid.
4. Suppose that two firms, Firm i $(i = 1, 2)$, are developing new products. When a firm succeeds in developing a new product faster than the other during $t \in [0, T]$, it will earn a market reward of R_1. If it succeeds later than the competitor, it will still earn a market reward R_2, smaller than R_1. Each firm has to invest in scientific knowledge k_i to win the game and the probability of innovation success of Firm i by t is $F_i(k_i(t)) = 1 - e^{-\lambda k_i(t)}$. Firm i's knowledge stock increases with the firm's effort $u_i(t)$, $\dot{k}_i(t) = u_i(t)$, and the initial knowledge stock of Firm i is $k_i(0) = k_{i0}$. The cost of innovation effort at t is given by $cu_i(t)^2$. We assume that the discounting effect can be disregarded. Formulate the differential game model of this problem. Construct the model for this innovation game for the planning period $[0, T]$.

Appendix 1: An Example of Differential Games Problem

In this section, we demonstrate how to obtain a feedback solution for a differential games problem, utilizing the HJB equation.

Feedback Nash Equilibrium

Consider the general objective function as follows:

$$V^i = \int_0^T \left((a_i x_i - a_j x_j) - e_i u_i^2 + e_j u_j^2 \right) dt + H\left(a_i x_i(T) - a_j x_j(T)\right),\ i \neq j \text{ and } i,j = 1,2 \tag{25}$$

More specifically, Player 1's decision problem writes:

$$\mathit{Maximize} \int_0^T ((a_1 x_1 - a_2 x_2) - e_1 u_1^2 + e_2 u_2^2) dt + H(a_1 x_1(T) - a_2 x_2(T)) \tag{26}$$

$$\begin{aligned} \mathit{Subject\ to}\, x_1^{'} &= b_1 u_1 \\ x_2^{'} &= b_2 u_2 \\ x_1(0) = x_2(0) &= 0\, given. \end{aligned}$$

Similarly, we define the decision problem for Player 2 as

$$\mathit{Maximize} \int_0^T ((-a_1 x_1 + a_2 x_2) + e_1 u_1^2 - e_2 u_2^2) dt + H(-a_1 x_1(T) + a_2 x_2(T)) \tag{27}$$

$$\begin{aligned} \mathit{Subject\ to}\, x_1^{'} &= b_1 u_1 \\ x_2^{'} &= b_2 u_2 \\ x_1(0) = x_2(0) &= 0\, given. \end{aligned}$$

Then, we solve the problem as follows.
First, define the HJB equations for the players:

$$-J_t^1 = max\left((a_1 x_1 - a_2 x_2) - e_1 u_1^2 + e_2 \bar{u}_2^2 + J_{x_1}^1 (b_1 u_1) + J_{x_2}^1 (b_2 \bar{u}_2) \right) \tag{28}$$

$$-J_t^2 = max\left((-a_1 x_1 + a_2 x_2) + e_1 \bar{u}_1^2 - e_2 u_2^2 + J_{x_1}^2 (b_1 \bar{u}_1) + J_{x_2}^2 (b_2 u_2) \right) \tag{29}$$

From (28) and (29), obtain the optimal controls:

$$-2e_1u_1 + b_1J^1_{x_1} = 0 \quad \therefore u_1^* = \frac{b_1}{2e_1}J^1_{x_1} \tag{30}$$

$$-2e_2u_2 + b_2J^2_{x_2} = 0 \quad \therefore u_2^* = \frac{b_2}{2e_2}J^2_{x_2} \tag{31}$$

Putting (30) and (31) in (28), we get:

$$\begin{aligned} -J^1_t &= (a_1x_1 - a_2x_2) - e_1\left(\frac{b_1^2}{4e_1^2}\left(J^1_{x_1}\right)^2\right) + e_2\left(\frac{b_2^2}{4e_2^2}\left(J^2_{x_2}\right)^2\right) + J^1_{x_1}b_1\left(\frac{b_1}{2e_1}J^1_{x_1}\right) + J^1_{x_2}b_2\left(\frac{b_2}{2e_2}J^2_{x_2}\right) \\ &= (a_1x_1 - a_2x_2) + \frac{b_1^2}{4e_1}\left(J^1_{x_1}\right)^2 + \frac{b_2^2}{4e_2}\left(J^2_{x_2}\right)^2 + \frac{b_2^2}{2e_2}\left(J^1_{x_2}J^2_{x_2}\right) \end{aligned} \tag{32}$$

From the structure of (25), we suggest

$$J^i = K(t)\left(a_ix_i - a_jx_j\right) + Q(t),\ i \neq j \text{ and } i,j = 1,2. \tag{33}$$

From (33), we obtain from (34) to (37):

$$J^1_t = K(t)'(a_1x_1 - a_2x_2) + Q(t)' \tag{34}$$

$$J^1_{x_1} = a_1K(t) \tag{35}$$

$$J^1_{x_2} = -a_2K(t) \tag{36}$$

$$J^2_{x_2} = a_2K(t) \tag{37}$$

By plugging (34)–(37) into (38), we have (39):

$$-J^1_t = (a_1x_1 - a_2x_2) + \frac{b_1^2}{4e_1}\left(J^1_{x_1}\right)^2 + \frac{b_2^2}{4e_2}\left(J^2_{x_2}\right)^2 + \frac{b_2^2}{2e_2}\left(J^1_{x_2}J^2_{x_2}\right) \tag{38}$$

$$\begin{aligned} -K(t)'(a_1x_1 - a_2x_2) - Q(t)' &= (a_1x_1 - a_2x_2) + \frac{b_1^2}{4e_1}(a_1K(t))^2 + \frac{b_2^2}{4e_2}(a_2K(t))^2 \\ &\quad + \frac{b_2^2}{2e_2}(-a_2K(t))(a_2K(t)) \end{aligned} \tag{39}$$

We rearrange (39) to get (40):

$$\therefore \left[\frac{K(t)^2}{4}\left(\frac{a_1^2b_1^2}{e_1}-\frac{a_2^2b_2^2}{e_2}\right)+Q(t)'\right]+(a_1x_1-a_2x_2)\left(1+K(t)'\right)=0 \tag{40}$$

To make (40) valid, we have to have:

$$1+K(t)'=0 \text{ and } \frac{K(t)^2}{4}\left(\frac{a_1^2b_1^2}{e_1}-\frac{a_2^2b_2^2}{e_2}\right)+Q(t)'=0 \tag{41}$$

The first part of (41) leads us to determine (42):

$$K(t)'=-1,\ K(t)=-t+c_1,\ \text{where } c_1 \text{ is a constant}$$
$$K(T)=-T+c_1=H,\ \text{therefore } c_1=T+H$$

$$K(t)=-t+T+H \tag{42}$$

Using (42) in (35) and (37), we derive:

$$J^1_{x_1}=a_1K(t)=a_1(-t+T+H);\ J^2_{x_2}=a_2K(t)=a_2(-t+T+H) \tag{43}$$

With (43), we determine the optimal control variables as in (44) and (45):

$$u_1^*=\frac{b_1}{2e_1}J^1_{x_1}=\frac{a_1b_1}{2e_1}(-t+T+H) \tag{44}$$

$$u_2^*=\frac{b_2}{2e_2}J^2_{x_2}=\frac{a_2b_2}{2e_2}(-t+T+H) \tag{45}$$

Using (44) and (45), we derive the optimal state variables as in (46) and (47):

$$x_1'=b_1u_1=b_1\frac{a_1b_1}{2e_1}(-t+T+H)=\frac{a_1b_1^2}{2e_1}(-t+T+H)$$
$$x_1(t)=-\frac{a_1b_1^2}{4e_1}t^2+\frac{a_1b_1^2}{2e_1}(T+H)t+k_1,\ \text{where } k_1 \text{ is a constant}$$

$$x_1(0)=k_1=0 \therefore x_1(t)=-\frac{a_1b_1^2}{4e_1}t^2+\frac{a_1b_1^2}{2e_1}(T+H)t \tag{46}$$

$$x_2' = b_2u_2 = b_2\frac{a_2b_2}{2e_2}(-t+T+H) = \frac{a_2b_2^2}{2e_2}(-t+T+H)$$

$$x_2(t) = -\frac{a_2b_2^2}{4e_2}t^2 + \frac{a_2b_2^2}{2e_2}(T+H)t + k_2, \text{ where } k_2 \text{ is a constant}$$

$$x_2(0) = k_2 = 0 \therefore x_2(t) = -\frac{a_2b_2^2}{4e_2}t^2 + \frac{a_2b_2^2}{2e_2}(T+H)t \tag{47}$$

Using (41) and (42), we determine $Q(t)$ as in (48):

$$Q(t)' = -\frac{K(t)^2}{4}\left(\frac{a_1^2b_1^2}{e_1} - \frac{a_2^2b_2^2}{e_2}\right) = -\frac{(-t+T+H)^2}{4}\left(\frac{a_1^2b_1^2}{e_1} - \frac{a_2^2b_2^2}{e_2}\right)$$

$$Q(t) = \frac{(-t+T+H)^3}{12}\left(\frac{a_1^2b_1^2}{e_1} - \frac{a_2^2b_2^2}{e_2}\right) + c_2, \text{ where } c_2 \text{ is a constant}$$

$$Q(T) = \frac{H^3}{12}\left(\frac{a_1^2b_1^2}{e_1} - \frac{a_2^2b_2^2}{e_2}\right) + c_2 = 0$$

$$\therefore Q(t) = \frac{(-t+T+H)^3 - H^3}{12}\left(\frac{a_1^2b_1^2}{e_1} - \frac{a_2^2b_2^2}{e_2}\right) \tag{48}$$

Verification of Value Function: To verify the optimal solution, we need to confirm that we have indeed determined a right value function in (33). We do this in two ways, i.e., calculating the optimal total profit by utilizing the objective function in (26) and the value function in (33).

A. Using the objective function in (26)

$$\begin{aligned}
V^1 &= \int_0^T ((a_1x_1 - a_2x_2) - e_1u_1^2 + e_2u_2^2)dt + H(a_1x_1(T) - a_2x_2(T)) \\
&= \int_0^T \left\{\left[a_1\left(-\frac{a_1b_1^2}{4e_1}t^2 + \frac{a_1b_1^2}{2e_1}(T+H)t\right) - a_2\left(-\frac{a_2b_2^2}{4e_2}t^2 + \frac{a_2b_2^2}{2e_2}(T+H)t\right)\right] - e_1\left[\frac{a_1b_1}{2e_1}(-t\right.\right. \\
&\quad \left.\left. +T+H)\right]^2 + e_2\left[\frac{a_2b_2}{2e_2}(-t+T+H)\right]^2\right\}dt \\
&\quad + H\left\{a_1\left[-\frac{a_1b_1^2}{4e_1}T^2 + \frac{a_1b_1^2}{2e_1}(T+H)T\right] - a_2\left[-\frac{a_2b_2^2}{4e_2}T^2 + \frac{a_2b_2^2}{2e_2}(T+H)T\right]\right\} \\
&= \frac{T(3H^2+3HT+T^2)}{12}\left(\frac{a_1^2b_1^2}{e_1} - \frac{a_2^2b_2^2}{e_2}\right) = \frac{(T+H)^3 - H^3}{12}\left(\frac{a_1^2b_1^2}{e_1} - \frac{a_2^2b_2^2}{e_2}\right).
\end{aligned} \tag{49}$$

B. Using the value function in (33)

$$J^1(0,x_1(0),x_2(0)) = K(0)(a_1x_1(0) - a_2x_2(0)) + Q(0) = K(0)(a_1 \times 0 - a_2 \times 0)$$
$$+ Q(0) = Q(0) = \frac{(T+H)^3 - H^3}{12}\left(\frac{a_1^2b_1^2}{e_1} - \frac{a_2^2b_2^2}{e_2}\right) \tag{50}$$

Since (49) is the same as (50), we know that the value function is correct and we have obtained an optimal solution.

Open-Loop Nash Equilibrium

Now let's find an open-loop Nash equilibrium solution for the differential games problem. Applying the maximum principle, we obtain optimal controls in (57) and (58) and also optimal state variables in (59) and (60).

$$H^1 = (a_1x_1 - a_2x_2) - e_1u_1^2 + e_2u_2^2 + \lambda_1^1 b_1 u_1 + \lambda_2^1 b_2 u_2 \tag{51}$$

$$H^2 = (-a_1x_1 + a_2x_2) + e_1u_1^2 - e_2u_2^2 + \lambda_1^2 b_1 u_1 + \lambda_2^2 b_2 u_2 \tag{52}$$

$$\lambda_1^{1\prime} = -H_{x_1}^1 = -a_1,\ \lambda_1^1(T) = a_1H \quad \therefore \lambda_1^1(t) = a_1(-t+T+H) \tag{53}$$

$$\lambda_2^{1\prime} = -H_{x_2}^1 = a_2,\ \lambda_2^1(T) = -a_2H \quad \therefore \lambda_2^1(t) = a_2(t-T-H) \tag{54}$$

$$\lambda_1^{2\prime} = -H_{x_1}^2 = a_1,\ \lambda_1^2(T) = -a_1H \quad \therefore \lambda_1^2(t) = a_1(t-T-H) \tag{55}$$

$$\lambda_2^{2\prime} = -H_{x_2}^2 = -a_2,\ \lambda_2^2(T) = a_2H \quad \therefore \lambda_2^2(t) = a_2(-t+T+H) \tag{56}$$

$$-2e_1u_1 + b_1\lambda_1^1 = 0 \quad \therefore u_1^* = \frac{b_1}{2e_1}\lambda_1^1 = \frac{a_1b_1}{2e_1}(-t+T+H) \tag{57}$$

$$-2e_2u_2 + b_2\lambda_2^2 = 0 \quad \therefore u_2^* = \frac{b_2}{2e_2}\lambda_2^2 = \frac{a_2b_2}{2e_2}(-t+T+H) \tag{58}$$

$$\begin{aligned} x_1' &= b_1u_1 = \frac{a_1b_1^2}{2e_1}(-t+T+H),\ x_1(0) = 0 \\ \therefore x_1(t) &= -\frac{a_1b_1^2}{4e_1}t^2 + \frac{a_1b_1^2}{2e_1}(T+H)t \end{aligned} \tag{59}$$

$$\begin{aligned} x_2' &= b_2 u_2 = \frac{a_2 b_2^2}{2e_2}(-t+T+H),\ x_2(0)=0 \\ \therefore x_2(t) &= -\frac{a_2 b_2^2}{4e_2}t^2 + \frac{a_2 b_2^2}{2e_2}(T+H)t \end{aligned} \tag{60}$$

For this particular differential games problem, we confirm that the open-loop Nash equilibrium solution is equivalent to the feedback Nash equilibrium solution.

Numerical Example

In order to visualize the optimal dynamics, we conduct a numerical analysis, using the parameter values in Table 5.

Figure 10 shows the optimal dynamics of the objective function and also those of the state variables. Figure 11 depicts the optimal dynamics of the control variables.

Table 5 Parameter values

a_1	b_1	a_2	b_2	T	H	e_1	e_2
8	4	5	8	50	0.1	1	1.56

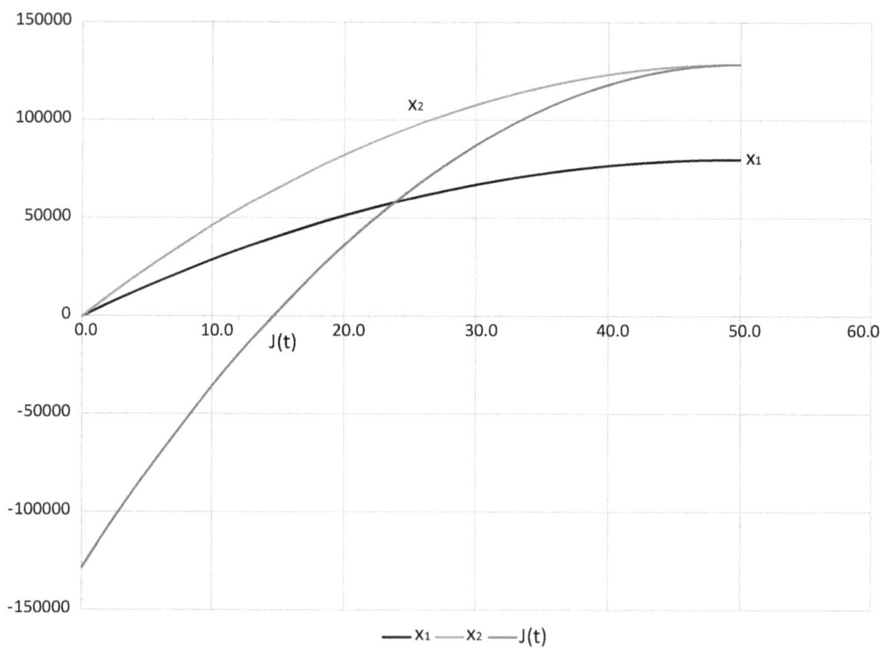

Fig. 10 Optimal dynamics of objective function and state variables

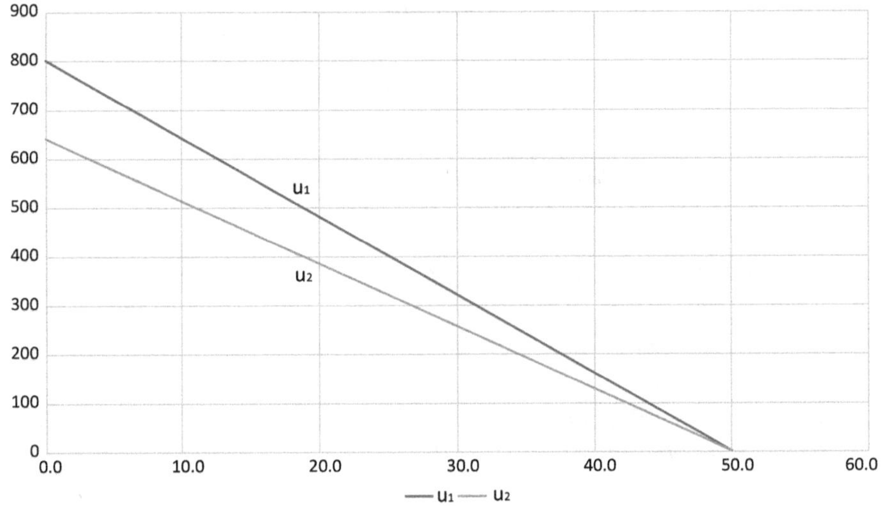

Fig. 11 Optimal dynamics of control variables

Appendix 2

Derivation of Theorem 1

Using the information in (19), we can reformulate each element of (17) as

(i) $J_t^1 = \dot{h}_1(t)e^{-\lambda(x_1+x_2)} + \dot{l}_1(t)e^{-\lambda x_2}$

(ii) $$\left(J_{x_1}^1\right)^2 \frac{3\alpha_1^2}{2K_1} t^{-n_1} e^{rt} e^{\lambda(x_1+x_2)} = \left(-\lambda h_1(t)e^{-\lambda(x_1+x_2)}\right)^2 \frac{3\alpha_1^2}{2K_1} t^{-n_1} e^{rt} e^{\lambda(x_1+x_2)}$$
$$= \lambda^2 \frac{3\alpha_1^2}{2K_1} t^{-n_1} e^{rt} e^{\lambda(x_1+x_2)} h_1^2$$

(iii) $$J_{x_2}^1 J_{x_2}^2 \frac{\alpha_2^2}{K_2} t^{-n_2} e^{rt} e^{\lambda(x_1+x_2)}$$
$$= \left(-\lambda h_1(t)e^{-\lambda(x_1+x_2)} - \lambda l_1(t)e^{-\lambda x_2}\right)\left(-\lambda h_2(t)e^{-\lambda(x_1+x_2)}\right) \frac{\alpha_2^2}{K_2} t^{-n_2} e^{rt} e^{\lambda(x_1+x_2)}$$
$$= \lambda^2 \frac{\alpha_2^2}{K_2} t^{-n_2} e^{rt} e^{-\lambda(x_1+x_2)} h_1 h_2 + \lambda^2 \frac{\alpha_2^2}{K_2} t^{-n_2} e^{rt} e^{-\lambda x_2} l_1 h_2$$

(iv) $$J_{x_2}^2 \frac{\alpha_2^2}{K_2} R_0 t^{m-n_2} e^{rt} \lambda\left(e^{\lambda x_1} - 1\right) = -\lambda h_2(t) e^{-\lambda(x_1+x_2)} \frac{\alpha_2^2}{K_2} R_0 t^{m-n_2} e^{rt} \lambda\left(e^{\lambda x_1} - 1\right) =$$
$$-\lambda^2 \frac{\alpha_2^2}{K_2} R_0 t^{m-n_2} e^{rt} e^{-\lambda x_2} h_2 + \lambda^2 \frac{\alpha_2^2}{K_2} R_0 t^{m-n_2} e^{rt} e^{-\lambda(x_1+x_2)} h_2$$

(v) $mR_0 t^{m-1}\left(1 - e^{-\lambda x_1}\right)e^{-\lambda x_2} = mR_0 t^{m-1} e^{-\lambda x_2} - mR_0 t^{m-1} e^{-\lambda(x_1+x_2)}$

By putting (i)–(v) into (17) and rearranging, we obtain:

$$E_1(t)e^{-\lambda(x_1+x_2)} + E_2(t)e^{-\lambda x_2} = 0, \tag{61}$$

where
$E_1(t) = \dot{h}_1 + \lambda^2 e^{rt}\left(\frac{3\alpha_1^2}{2K_1}t^{-n_1}h_1^2 + \frac{\alpha_2^2}{K_2}t^{-n_2}h_1h_2\right) + \lambda^2 e^{rt}\frac{\alpha_2^2}{K_2}R_0t^{m-n_2}h_2 + mR_0t^{m-1}$ and

$$\begin{aligned} E_2(t) &= \dot{l}_1 + \lambda^2 e^{rt}\frac{\alpha_2^2}{K_2}t^{-n_2}l_1h_2 - \lambda^2 e^{rt}\frac{\alpha_2^2}{K_2}R_0t^{m-n_2}h_2 - mR_0t^{m-1} \\ &= \dot{l}_1 + \lambda^2 e^{rt}\frac{\alpha_2^2}{K_2}t^{-n_2}(l_1 - R_0t^m)h_2 - mR_0t^{m-1}. \end{aligned}$$

In order to satisfy (61), both $E_1(t)$ and $E_2(t)$ must be zero. In addition, we can confirm that $l_1(t) = R_0t^m$ makes $E_2(t) = 0$.

Derivation of Corollary 1

Following steps similar with those for Firm 1, we can propose

$$J^2(t, x_1, x_2) = h_2(t)e^{-\lambda(x_1+x_2)} + l_2(t)e^{-\lambda x_1}.$$

Using (19) and rearranging appropriately, we have the following condition for optimality: $G_1(t)e^{-\lambda(x_1+x_2)} + G_2(t)e^{-\lambda x_1} = 0$, where

$$G_1(t) = \dot{h}_2 + \lambda^2 e^{rt}\left(\frac{3\alpha_2^2}{2K_2}t^{-n_2}h_2^2 + \frac{\alpha_1^2}{K_1}t^{-n_1}h_1h_2\right) + \lambda^2 e^{rt}\frac{\alpha_1^2}{K_1}R_0t^{m-n_1}h_1 + mR_0t^{m-1} \text{ and}$$

$$\begin{aligned} G_2(t) &= \dot{l}_2 + \lambda^2 e^{rt}\frac{\alpha_1^2}{K_1}t^{-n_1}l_2h_1 - \lambda^2 e^{rt}\frac{\alpha_1^2}{K_1}R_0t^{m-n_1}h_1 - mR_0t^{m-1} \\ &= \dot{l}_2 + \lambda^2 e^{rt}\frac{\alpha_1^2}{K_1}t^{-n_1}(l_2 - R_0t^m)h_1 - mR_0t^{m-1}. \end{aligned}$$

Again, it must be that $G_1(t) = G_2(t) = 0$. We can also show that $l_2 = R_0t^m$ makes $G_2(t) = 0$.

Derivation of Theorem 3

Consider for $i = 1$, using HJB equation for an infinity case:

$$rJ = \max_{u_1}\left\{ (x_1 + x_2)[a_i - (x_1 + x_2)] + s_i x_i [b_i - x_i] - \rho u_i - \frac{1}{2}u_i^2 + J_{x_1}(u_1 - \delta x_1) \right.$$
$$\left. + J_{x_2}(u_2 - \delta x_2) \right\} \tag{62}$$

Thus, we obtain: $u_1 = J_{x_1} - \rho$, $u_2 = J^2_{x_2} - \rho$.
We suggest the following value functions:

$$J = \frac{a}{2}(x_1 + x_2)^2 + \beta_1(x_1 + x_2) + \frac{\theta}{2}x_1^2 + \phi_1 x_1 + \gamma_1 \text{ for } i = 1\text{ , and}$$
$$J^2 = \frac{a}{2}(x_1 + x_2)^2 + \beta_2(x_1 + x_2) + \frac{\theta}{2}x_2^2 + \phi_2 x_2 + \gamma_2 \text{ for } i = 2.$$

Take appropriate partial differentiations and plug them into (62):

$$J_{x_1} = \alpha(x_1 + x_2) + \beta_1 + \theta x_1 + \phi_1$$
$$J_{x_2} = \alpha(x_1 + x_2) + \beta_1$$
$$J^2_{x_2} = \alpha(x_1 + x_2) + \beta_2 + \theta x_2 + \phi_2$$
$$\therefore rJ = (x_1 + x_2)[a_1 - (x_1 + x_2)] + s_1 x_1 [b_1 - x_1] - \rho(J_{x_1} - \rho) - \frac{1}{2}(J_{x_1} - \rho)^2$$
$$+ J_{x_1}(J_{x_1} - \rho - \delta x_1) + J_{x_2}(J^2_{x_2} - \rho - \delta x_2)$$
$$\therefore rJ = (x_1 + x_2)[a_1 - (x_1 + x_2)] + s_1 x_1 [b_1 - x_1] + \frac{1}{2}(J_{x_1})^2 - \rho J_{X_1} - \delta x_1 J_{x_1} + \frac{1}{2}\rho^2$$
$$+ J_{x_2} J^2_{x_2} - \rho J_{x_2} - \delta x_2 J_{x_2}$$

For simplicity, we consider separate cases:

(1) $\frac{1}{2}(J_{x_1})^2 = \frac{1}{2}\left\{\alpha^2(x_1 + x_2)^2 + \theta^2 x_1^2 + (\beta_1 + \phi_1)^2 + 2\alpha\theta x_1(x_1 + x_2) + 2\alpha(\beta_1 + \phi_1)(x_1 + x_2) + 2\theta(\beta_1 + \phi_1)x_1\right\}$

(2) $\rho J_{x_1} = \alpha\rho(x_1 + x_2) + \theta\rho x_1 + \rho(\beta_1 + \phi_1)$

(3) $\delta x_1 J_{x_1} = \alpha\delta x_1(x_1 + x_2) + \theta\delta x_1^2 + \delta(\beta_1 + \phi_1)x_1$

(4) $J_{x_2} J^2_{x_2} = \alpha^2(x_1 + x_2)^2 + \alpha\theta x_2(x_1 + x_2) + \alpha(\beta_2 + \phi_2)(x_1 + x_2) + \alpha\beta_1(x_1 + x_2) + \theta\beta_1 x_2 + \beta_1(\beta_2 + \phi_2)$

(5) $\rho J_{x_2} = \alpha\rho(x_1 + x_2) + \rho\beta_1$

(6) $\delta x_2 J_{x_2} = \alpha\delta x_2(x_1 + x_2) + \delta\beta_1 x_2$

Now,

$$\begin{aligned}
&\therefore -r\frac{\alpha}{2}(x_1+x_2)^2 - r\beta_1(x_1+x_2) - r\frac{\theta}{2}x_1^2 - r\phi_1 x_1 - r\gamma_1 \\
&+a_1(x_1+x_2) - (x_1+x_2)^2 + s_1 b_1 x_1 - s_1 x_1^2 \\
&+\frac{1}{2}\alpha^2(x_1+x_2)^2 + \frac{1}{2}\theta^2 x_1^2 + \frac{1}{2}(\beta_1+\phi_1)^2 + \alpha\theta x_1(x_1+x_2) \\
&+\alpha(\beta_1+\phi_1)(x_1+x_2) + \theta(\beta_1+\phi_1)x_1 \\
&-\alpha\rho(x_1+x_2) - \theta\rho x_1 - \rho(\beta_1+\phi_1) \\
&-\alpha\delta x_1(x_1+x_2) - \theta\delta x_1^2 - \delta(\beta_1+\phi_1)x_1 + \frac{1}{2}\rho^2 \\
&+\alpha^2(x_1+x_2)^2 + a\theta x_2(x_1+x_2) + \alpha(\beta_2+\phi_2)(x_1+x_2) + \alpha\beta_1(x_1+x_2) + \theta\beta_1 x_2 \\
&+\beta_1(\beta_2+\phi_2) - \alpha\rho(x_1+x_2) - \rho\beta_1 - \alpha\delta x_2(x_1+x_2) - \delta\beta_1 x_2 = 0
\end{aligned}$$

$$\begin{aligned}
&\therefore\left\{-r\frac{a}{2} - 1 + \frac{3}{2}\alpha^2\right\}(x_1+x_2)^2 \\
&+\{-r\beta_1 + a_1 + \alpha(\beta_1+\phi_1) - \alpha\rho + \alpha(\beta_2+\phi_2) + \alpha\beta_1 - \alpha\rho\}(x_1+x_2) \\
&+\left\{-r\frac{\theta}{2} - s_1 + \frac{1}{2}\theta^2 - \theta\delta\right\}x_1^2 + \{-r\phi_1 + s_1 b_1 + \theta(\beta_1+\phi_1) - \theta\rho - \delta(\beta_1+\phi_1)\}x_1 \\
&+\{\alpha\theta - \alpha\delta\}x_1(x_1+x_2) + \{\alpha\theta - \alpha\delta\}x_2(x_1+x_2) + \{\theta\beta_1 - \delta\beta_1\}x_2 \\
&+\left\{-r\gamma_1 + \frac{1}{2}(\beta_1+\phi_1)^2 - \rho(\beta_1+\phi_1) + \frac{1}{2}\rho^2 + \beta_1(\beta_2+\phi_2) - \rho\beta_1\right\} = 0
\end{aligned}$$

$$\therefore Ax_1^2 + Bx_1x_2 + Cx_2^2 + Dx_1 + Ex_2 + F = 0$$

Therefore, we must have: $A = B = C = D = E = F = 0$, where

$$\begin{aligned}
A &= \left\{-r\frac{\alpha}{2} + 1 + \frac{3}{2}\alpha^2\right\} + \left\{-r\frac{\theta}{2} - s_1 + \frac{1}{2}\theta^2 - \theta\delta\right\} + \{\alpha\theta - \alpha\delta\} \\
B &= 2\left\{-r\frac{\alpha}{2} - 1 + \frac{3}{2}\alpha^2\right\} + \{\alpha\theta - \alpha\delta\} + \{\alpha\theta - \alpha\delta\} \\
C &= \left\{-r\frac{\alpha}{2} - 1 + \frac{3}{2}\alpha^2\right\} + \{\alpha\theta - \alpha\delta\} \\
D &= \{-r\beta_1 + a_1 + \alpha(2\beta_1 + \beta_2 + \phi_1 + \phi_2 - 2\rho)\} + \{-r\phi_1 + s_1 b_1 + (\theta - \delta)(\beta_1+\phi_1) - \theta\rho\} \\
E &= \{-r\beta_1 + a_1 + \alpha(2\beta_1 + \beta_2 + \phi_1 + \phi_2 - 2\rho)\} + \{\theta\beta_1 - \delta\beta_1\} \\
F &= -r\gamma_1 + \frac{1}{2}(\beta_1+\phi_1)^2 - \rho(\beta_1+\phi_1) + \frac{1}{2}\rho^2 + \beta_1(\beta_2+\phi_2) - \rho\beta_1.
\end{aligned}$$

From $A = C = 0$, it must be that $\left\{-r\frac{\theta}{2} - s_1 + \frac{1}{2}\theta^2 - \theta\delta\right\} = \frac{1}{2}\theta^2 - \left(\frac{r}{2} + \delta\right)\theta - s_1 = 0$.

Therefore, $\theta = \frac{(r+2\delta)\pm\sqrt{(r+2\delta)^2+8s_1}}{2}$. We also need to impose $\theta<0$, thus $\theta = \frac{(r+2\delta)-\sqrt{(r+2\delta)^2+8s_1}}{2}$.

From $B = C = 0$, $\left\{-r\frac{\alpha}{2} - 1 + \frac{3}{2}\alpha^2\right\} + \{\alpha\theta - \alpha\delta\} = \frac{3}{2}\alpha^2 - \left(\frac{r}{2} + \delta - \theta\right)\alpha - 1 = 0$, and $3\alpha^2 - (r + 2\delta - 2\theta)\alpha - 2 = 0$. Therefore, $\alpha = \frac{(r+2\delta-2\theta)\pm\sqrt{(r+2\delta-2\theta)^2+24}}{6}$. Due to the condition $\alpha<0$, $\alpha = \frac{(r+2\delta-2\theta)-\sqrt{(r+2\delta-2\theta)^2+24}}{6}$.

Taking the similar steps for $i = 2$, we obtain the following:

$$D^2 = \{-r\beta_2 + a_2 + \alpha(2\beta_2 + \beta_1 + \phi_1 + \phi_2 - 2\rho)\} + \{-r\phi_2 + s_2 b_2 + (\theta - \delta)(\beta_2 + \phi_2) - \theta\rho\}$$
$$E^2 = \{-r\beta_2 + a_2 + \alpha(2\beta_2 + \beta_1 + \phi_1 + \phi_2 - 2\rho)\} + \{\theta\beta_2 - \delta\beta_2\}$$
$$F^2 = -r\gamma_2 + \frac{1}{2}(\beta_2 + \phi_2)^2 - \rho(\beta_2 + \phi_2) + \frac{1}{2}\rho^2 + \beta_2(\beta_1 + \phi_1) - \rho\beta_2$$

Now, we have 4 equations, D, E, D^2, and E^2, and 4 variables, β_1, β_2, ϕ_1, and ϕ_2, to determine.

$$K\begin{bmatrix}\beta_1\\ \beta_2\\ \phi_1\\ \phi_2\end{bmatrix} = L, \text{ where } K$$
$$\equiv \begin{bmatrix} 2\alpha+\theta-r-\delta & \alpha & \alpha+\theta-r-\delta & \alpha \\ 2\alpha+\theta-r-\delta & \alpha & \alpha & \alpha \\ \alpha & 2\alpha+\theta-r-\delta & \alpha & \alpha+\theta-r-\delta \\ \alpha & 2\alpha+\theta-r-\delta & \alpha & \alpha \end{bmatrix}$$

$$\text{and } L \equiv -\begin{bmatrix} a_1 - 2\alpha\rho + s_1 b_1 - \theta\rho \\ a_1 - 2\alpha\rho \\ a_2 - 2\alpha\rho + s_2 b_2 - \theta\rho \\ a_2 - 2\alpha\rho \end{bmatrix}$$

$$\times\left[\begin{array}{cccc|c} 2\alpha+\theta-r-\delta & \alpha & \alpha+\theta-r-\delta & \alpha & 2\alpha\rho+\theta\rho-a_1-s_1b_1 \\ 2\alpha+\theta-r-\delta & \alpha & \alpha & \alpha & 2\alpha\rho-a_1 \\ \alpha & 2\alpha+\theta-r-\delta & \alpha & \alpha+\theta-r-\delta & 2\alpha\rho+\theta\rho-a_2-s_2b_2 \\ \alpha & 2\alpha+\theta-r-\delta & \alpha & \alpha & 2\alpha\rho-a_2 \end{array}\right]$$

$$\times\left[\begin{array}{cccc|c} 2\alpha+\theta-r-\delta & \alpha & \alpha+\theta-r-\delta & \alpha & 2\alpha\rho+\theta\rho-a_1-s_1b_1 \\ 0 & 0 & r+\delta-\theta & 0 & s_1b_1-\theta\rho \\ \alpha & 2\alpha+\theta-r-\delta & \alpha & \alpha+\theta-r-\delta & 2\alpha\rho+\theta\rho-a_2-s_2b_2 \\ 0 & 0 & 0 & r+\delta-\theta & s_2b_2-\theta\rho \end{array}\right]$$

$\therefore \phi_1 = \frac{s_1b_1-\theta\rho}{r+\delta-\theta}$ and $\phi_2 = \frac{s_2b_2-\theta\rho}{r+\delta-\theta}$

$$(2\alpha+\theta-r-\delta)\beta_1+\alpha\beta_2+\frac{(\alpha+\theta-r-\delta)(s_1b_1-\theta\rho)}{r+\delta-\theta}+\frac{\alpha(s_2b_2-\theta\rho)}{r+\delta-\theta}=2\alpha\rho+\theta\rho-a_1-s_1b_1$$
$$\alpha\beta_1+(2\alpha+\theta-r-\delta)\beta_2+\frac{\alpha(s_1b_1-\theta\rho)}{r+\delta-\theta}+\frac{(\alpha+\theta-r-\delta)(s_2b_2-\theta\rho)}{r+\delta-\theta}=2\alpha\rho+\theta\rho-a_2-s_2b_2$$

Let's define $W_i=\frac{\alpha(2r\rho+2\delta\rho-s_1b_1-s_2b_2)}{r+\delta-\theta}-a_i$.
Then, $(2\alpha+\theta-r-\delta)\beta_1+\alpha\beta_2=W_1$ and $\alpha\beta_1+(2\alpha+\theta-r-\delta)\beta_2=W_2$.

$$\therefore \beta_1=\frac{2\alpha+\theta-r-\delta}{(2\alpha+\theta-r-\delta)^2-\alpha^2}W_1+\frac{-\alpha}{(2\alpha+\theta-r-\delta)^2-\alpha^2}W_2 \text{ and}$$
$$\beta_2=\frac{-\alpha}{(2\alpha+\theta-r-\delta)^2-\alpha^2}W_1+\frac{2\alpha+\theta-r-\delta}{(2\alpha+\theta-r-\delta)^2-\alpha^2}W_2$$

Finally, from F, $\gamma_1=\frac{1}{r}\left[\frac{1}{2}(\beta_1+\phi_1)^2-\rho(\beta_1+\phi_1)+\frac{1}{2}\rho^2+\beta_1(\beta_2+\phi_2)-\rho\beta_1\right]$.
Likewise, from F^2, $\gamma_2=\frac{1}{r}\left[\frac{1}{2}(\beta_2+\phi_2)^2-\rho(\beta_2+\phi_2)+\frac{1}{2}\rho^2+\beta_2(\beta_1+\phi_1)-\rho\beta_2\right]$.

$$\text{Now, } u_1=J_{x_1}-\rho=\alpha(x_1+x_2)+\theta x_1+\beta_1+\phi_1-\rho \text{ and}$$
$$u_2=J^2_{x_2}-\rho=\alpha(x_1+x_2)+\theta x_2+\beta_2+\phi_2-\rho.$$

In order to get exact expressions, we need to do the following:

$$\begin{aligned}\dot{x}_1&=u_1-\delta x_1=\alpha(x_1+x_2)+(\theta-\delta)x_1+\beta_1+\phi_1-\rho\\&=(\alpha+\theta-\delta)x_1+\alpha x_2+\beta_1+\phi_1-\rho\end{aligned}$$

$$\begin{aligned}\dot{x}_2&=u_2-\delta x_2=\alpha(x_1+x_2)+(\theta-\delta)x_2+\beta_2+\phi_2-\rho\\&=\alpha x_1+(\alpha+\theta-\delta)x_2+\beta_2+\phi_2-\rho.\end{aligned}$$

Homogeneous equations:

$$\ddot{x}_1-2(\alpha+\theta-\delta)\dot{x}_1+\left[(\alpha+\theta-\delta)^2-\alpha^2\right]=0$$
$$y^2-2(\alpha+\theta-\delta)y+\left[(\alpha+\theta-\delta)^2-\alpha^2\right]=0$$
$$y=(\alpha+\theta-\delta)\pm\alpha.$$

Therefore, $x_1=c_1e^{(2\alpha+\theta-\delta)t}+c_2e^{(\theta-\delta)t}+N_1$.
Since $x_2=\frac{1}{\alpha}[\dot{x}_1-(\alpha+\theta-\delta)x_1]$, $x_2=c_1e^{(2\alpha+\theta-\delta)t}-c_2e^{(\theta-\delta)t}+N_2$.

Now, a particular solution:

$$0 = (\alpha + \theta - \delta)x_1 + \alpha x_2 + \beta_1 + \phi_1 - \rho$$
$$0 = \alpha x_1 + (\alpha + \theta - \delta)x_2 + \beta_2 + \phi_2 - \rho$$
$$x_2 = \frac{-\alpha x_1 - (\beta_2 + \phi_2 - \rho)}{\alpha + \theta - \delta}$$
$$(\alpha + \theta - \delta)x_1 - \frac{\alpha^2}{\alpha + \theta - \delta}x_1 - \frac{\alpha(\beta_2 + \phi_2 - \rho)}{\alpha + \theta - \delta} + \beta_1 + \phi_1 - \rho = 0$$
$$(\alpha + \theta - \delta)^2 x_1 - \alpha^2 x_1 - \alpha(\beta_2 + \phi_2 - \rho) + (\alpha + \theta - \delta)(\beta_1 + \phi_1 - \rho) = 0$$
$$x_1 = \frac{\alpha(\beta_2 + \phi_2 - \rho) - (\alpha + \theta - \delta)(\beta_1 + \phi_1 - \rho)}{(\alpha + \theta - \delta)^2 - \alpha^2}$$
$$x_2 = \frac{-\alpha}{\alpha + \theta - \delta}\left(\frac{\alpha(\beta_2 + \phi_2 - \rho) - (\alpha + \theta - \delta)(\beta_1 + \phi_1 - \rho)}{(\alpha + \theta - \delta)^2 - \alpha^2}\right) - \frac{\beta_2 + \phi_2 - \rho}{\alpha + \theta - \delta}$$
$$x_2 = \frac{\alpha(\beta_1 + \phi_1 - \rho) - (\alpha + \theta - \delta)(\beta_2 + \phi_2 - \rho)}{(\alpha + \theta - \delta)^2 - \alpha^2}$$

Therefore,

$$x_1 = c_1 e^{(2\alpha + \theta - \delta)t} + c_2 e^{(\theta - \delta)t} + \frac{\alpha(\beta_2 + \phi_2 - \rho) - (\alpha + \theta - \delta)(\beta_1 + \phi_1 - \rho)}{(\alpha + \theta - \delta)^2 - \alpha^2}$$
$$x_2 = c_1 e^{(2\alpha + \theta - \delta)t} - c_2 e^{(\theta - \delta)t} + \frac{\alpha(\beta_1 + \phi_1 - \rho) - (\alpha + \theta - \delta)(\beta_2 + \phi_2 - \rho)}{(\alpha + \theta - \delta)^2 - \alpha^2}$$

To determine c_1 and c_2, assume $x_1(0) = x_{10}$ and $x_2(0) = \mathrm{x}_{20}$. (Further suppose $x_1(0) = x_2(0) = 0$ to calculate c_1 and c_2.).

Derivation of Theorem 4

Now, consider a perfect coordination (i.e., collusive) case:

$$J = \int_0^\infty e^{-rt}\{(x_1 + x_2)[a_1 + a_2 - 2(x_1 + x_2)] + s_1 x_1[b_1 - x_1] + s_2 x_2[b_2 - x_2]$$
$$-\rho(u_1 + u_2) - \frac{1}{2}\left(u_1^2 + u_2^2\right)\Big\}dt$$

$$\dot{x}_1 = u_1 - \delta x_1$$
$$\dot{x}_2 = u_2 - \delta x_2$$

Current value Hamiltonian is:

$$H = (x_1 + x_2)[a_1 + a_2 - 2(x_1 + x_2)] + s_1 x_1[b_1 - x_1] + s_2 x_2[b_2 - x_2] - \rho(u_1 + u_2) \\ - \frac{1}{2}\left(u_1^2 + u_2^2\right) + \lambda_1(u_1 - \delta x_1) + \lambda_2(u_2 - \delta x_2)$$

Necessary conditions:

$$\frac{\partial H}{\partial u_1} = -\rho - u_1 + \lambda_1 = 0,\ u_1 = \lambda_1 - \rho;\ \frac{\partial H}{\partial u_2} = -\rho - u_2 + \lambda_2 = 0,\ u_2 = \lambda_2 - \rho$$

$$\dot{\lambda}_1 = r\lambda_1 - \frac{\partial H}{\partial x_1} = (r + \delta)\lambda_1 + (2s_1 + 4)x_1 + 4x_2 - (a_1 + a_2 + s_1 b_1).$$

Likewise, $\dot{\lambda}_2 = r\lambda_2 - \frac{\partial H}{\partial x_2} = (r + \delta)\lambda_2 + 4x_1 + (2s_2 + 4)x_2 - (a_1 + a_2 + s_2 b_2)$.
Now, we have

$$\begin{aligned}
\dot{x}_1 &= -\delta x_1 + \lambda_1 - \rho \\
\dot{x}_2 &= -\delta x_2 + \lambda_2 - \rho \\
\dot{\lambda}_1 &= (2s_1 + 4)x_1 + 4x_2 + (r + \delta)\lambda_1 - (a_1 + a_2 + s_1 b_1) \\
\dot{\lambda}_2 &= 4x_1 + (2s_2 + 4)x_2 + (r + \delta)\lambda_2 - (a_1 + a_2 + s_2 b_2)
\end{aligned}$$

For homogeneous systems,

$$\begin{bmatrix} -\delta & 0 & 1 & 0 \\ 0 & -\delta & 0 & 1 \\ 2s_1 + 4 & 4 & r + \delta & 0 \\ 4 & 2s_2 + 4 & 0 & r + \delta \end{bmatrix} \begin{bmatrix} x_1 \\ x_2 \\ \lambda_1 \\ \lambda_2 \end{bmatrix} \begin{vmatrix} -\delta - z & 0 & 1 & 0 \\ 0 & -\delta - z & 0 & 1 \\ 2s_1 + 4 & 4 & r + \delta - z & 0 \\ 4 & 2s_2 + 4 & 0 & r + \delta - z \end{vmatrix} = 0$$

$$(-\delta - z)\begin{vmatrix} -\delta - z & 0 & 1 \\ 4 & r + \delta - z & 0 \\ 2s_2 + 4 & 0 & r + \delta - z \end{vmatrix} + \begin{vmatrix} 0 & -\delta - z & 1 \\ 2s_1 + 4 & 4 & 0 \\ 4 & 2s_2 + 4 & r + \delta - z \end{vmatrix} = 0$$

$$(-\delta - z)\left\{(-\delta - z)(r + \delta - z)^2 - (2s_2 + 4)(r + \delta - z)\right\} + (\delta + z)(2s_1 + 4)(r + \delta - z) \\ + (2s_1 + 4)(2s_2 + 4) - 16 = 0$$

$$\{(\delta + z)(r + \delta - z)\}^2 + (2s_1 + 2s_2 + 8)\{(\delta + z)(r + \delta - z)\} + (2s_1 + 4)(2s_2 + 4) - 16 = 0$$

$$(\delta + z)(r + \delta - z) = \frac{-(2s_1 + 2s_2 + 8) \pm \sqrt{(2s_1 + 2s_2 + 8)^2 - 4\{(2s_1 + 4)(2s_2 + 4) - 16\}}}{2}$$

and we can determine 4 values of z.

For instance, suppose the case when $s_1 = s_2 = 1$.

$$[\{(\delta + z)(r + \delta - z)\} + 2][\{(\delta + z)(r + \delta - z)\} + 10] = 0.$$

Therefore, $(\delta + z)(r + \delta - z) = -2$ or $(\delta + z)(r + \delta - z) = -10$.
From $(\delta + z)(r + \delta - z) = -2$, we know $z = \frac{r \pm \sqrt{r^2 + 4\{\delta(r+\delta) + 2\}}}{2}$.
From $(\delta + z)(r + \delta - z) = -10$, we know $z = \frac{r \pm \sqrt{r^2 + 4\{\delta(r+\delta) + 10\}}}{2}$.
In order for the problem to have a solution, it must hold that $\lim_{t\to\infty} x_i(t)$ converges.

Therefore, out of four possible values, we can have only two, i.e., $z_1 = \frac{r - \sqrt{r^2 + 4\{\delta(r+\delta) + 2\}}}{2}$ and $z_3 = \frac{r - \sqrt{r^2 + 4\{\delta(r+\delta) + 10\}}}{2}$ by posing $A_2 = A_4 = B_2 = B_4 = C_2 = C_4 = D_2 = D_4 = 0$.
Now, we have the following system.

$$\begin{aligned}
x_1 &= A_1 e^{z_1 t} + A_3 e^{z_3 t} + Constant \\
x_2 &= B_1 e^{z_1 t} + B_3 e^{z_3 t} + Constant \\
\lambda_1 &= C_1 e^{z_1 t} + C_3 e^{z_3 t} + Constant \\
\lambda_2 &= D_1 e^{z_1 t} + D_3 e^{z_3 t} + Constant
\end{aligned}$$

In order to determine the constant terms, we can use the following set of relations.

$$0 = -\delta x_1 + \lambda_1 - \rho \tag{63}$$

$$0 = -\delta x_2 + \lambda_2 - \rho \tag{64}$$

$$\begin{aligned}
0 &= (2s_1 + 4)x_1 + 4x_2 + (r + \delta)\lambda_1 - (a_1 + a_2 + s_1 b_1) \\
0 &= 4x_1 + (2s_2 + 4)x_2 + (r + \delta)\lambda_2 - (a_1 + a_2 + s_2 b_2)
\end{aligned}$$

After some arrangements, we have the following:

$$\begin{aligned}
Const(x_1) &= \frac{4[\rho(r+\delta) - (a_1 + a_2 + s_2 b_2)] - [\rho(r+\delta) - (a_1 + a_2 + s_1 b_1)][\delta(r+\delta) + 2s_2 + 4]}{[\delta(r+\delta) + 2s_1 + 4][\delta(r+\delta) + 2s_2 + 4] - 16} \\
Const(x_2) &= \frac{4[\rho(r+\delta) - (a_1 + a_2 + s_1 b_1)] - [\rho(r+\delta) - (a_1 + a_2 + s_2 b_2)][\delta(r+\delta) + 2s_1 + 4]}{[\delta(r+\delta) + 2s_1 + 4][\delta(r+\delta) + 2s_2 + 4] - 16}.
\end{aligned}$$

Now we have:

$$\begin{aligned}
x_1 &= A_1 e^{z_1 t} + A_3 e^{z_3 t} + Const(x_1) \\
x_2 &= B_1 e^{z_1 t} + B_3 e^{z_3 t} + Const(x_2)
\end{aligned}$$

From (63) and (64), we know:

$$Const(\lambda_1) = \delta Const(x_1) + \rho \text{ and } Const(\lambda_2) = \delta Const(x_2) + \rho.$$
$$\lambda_1 = C_1 e^{z_1 t} + C_3 e^{z_3 t} + Const(\lambda_1)$$
$$\lambda_2 = D_1 e^{z_1 t} + D_3 e^{z_3 t} + Const(\lambda_2)$$

Now, how to determine A_1, A_3, B_1, B_3, C_1, C_3, D_1, D_3 ?
Since $\dot{x}_1 = -\delta x_1 + \lambda_1 - \rho$ and $\dot{x}_2 = -\delta x_2 + \lambda_2 - \rho$,

$$A_1 z_1 e^{z_1 t} + A_3 z_3 e^{z_3 t} = -\delta A_1 e^{z_1 t} - \delta A_3 e^{z_3 t} - \delta Const(x_1) + C_1 e^{z_1 t} + C_3 e^{z_3 t} + Const(\lambda_1) - \rho.$$

$$A_1(z_1 + \delta) = C_1,\ A_3(z_3 + \delta) = C_3,\ \delta Const(x_1) - Const(\lambda_1) + \rho = 0.$$

Likewise, $B_1(z_1 + \delta) = D_1$, $D_3(z_3 + \delta) = D_3$.

Therefore, we only need to determine A_1, A_3, B_1, B_3. It will be possible to determine exact values of these parameters, if specific initial conditions for the state variables are provided. For our purpose, we only need the steady-state values, which are represented by the constants in the state variables.

Derivation of Observation 1

$X_{1N} \le X_{1C}$ implies $\frac{2\rho(r+\delta-\alpha)-(a_1+a_2)}{(r+\delta-3\alpha)(2\alpha-\delta)} \le \frac{2[(a_1+a_2)-\rho(r+\delta)]}{\delta(r+\delta)+8}$.

That is, $\because (r + \delta - 3\alpha)(2\alpha - \delta) < 0$,

$$(\delta(r+\delta)+8)[2\rho(r+\delta-\alpha)-(a_1+a_2)] \ge 2[(a_1+a_2)-\rho(r+\delta)](r+\delta-3\alpha)(2\alpha-\delta) \tag{65}$$

Now,

$$\begin{aligned} RHS &= 2[(a_1+a_2)-\rho(r+\delta)](r+\delta-3\alpha)(2\alpha-\delta) \\ &= 2[(a_1+a_2)-\rho(r+\delta)]\left[-6\alpha^2+\alpha(2r+5\delta)-\delta(r+\delta)\right] \end{aligned}$$

and $3\alpha^2 = (r + 2\delta)\alpha + 2$ (see p. 261 in Dockner et al. (2000)).

$$\therefore RHS = 2[(a_1+a_2)-\rho(r+\delta)](\delta\alpha - 4 - \delta(r+\delta))$$

Then, from (65)

$$(\delta(r+\delta)+8)[2\rho(r+\delta)-(a_1+a_2)]-2\alpha\rho(\delta(r+\delta)+8)$$
$$\geq 2[(a_1+a_2)-\rho(r+\delta)](\delta\alpha-4-\delta(r+\delta))$$
$$(\delta(r+\delta)+8)[2\rho(r+\delta)-(a_1+a_2)]+2[(a_1+a_2)-\rho(r+\delta)]\delta(r+\delta)\geq 2[(a_1$$
$$+a_2)-[(\rho(r+\delta)](\delta\alpha-4)+2\alpha\rho(\delta(r+\delta)+8)$$
$$\delta(r+\delta)(a_1+a_2)+16\rho(r+\delta)-8(a_1+a_2)$$
$$\geq 2\delta\alpha(a_1+a_2)-8(a_1+a_2)+8\rho(r+\delta)+16\alpha\rho$$
$$\delta(r+\delta)(a_1+a_2)+8\rho(r+\delta)\geq 2\alpha\{8\rho+\delta(a_1+a_2)\}$$
$$(r+\delta)\{8\rho+\delta(a_1+a_2)\}\geq 2\alpha\{8\rho+\delta(a_1+a_2)\}$$

Therefore, $X_{1N}\leq X_{1C}$ *if and only if* $\frac{(r+\delta)}{2}\geq\alpha$.

Since α is negative for all r and δ, $\frac{(r+\delta)}{2}\geq\alpha$ holds always and so does $X_{1N}\leq X_{1C}$.

Appendix 3

See Appendix Tables 6 and 7.

Table 6 Industry level investment—common benefits only

A Markov perfect Nash equilibrium	A perfect coordination
$\lim_{t\to\infty} x(t) = -\frac{\beta_1+\beta_2-2\rho}{2\alpha-\delta}$ $\beta_i=\frac{1}{\Delta_m}\left[2\alpha^2\rho+(r+\delta)(a_i-2\alpha\rho)-\alpha(2a_i-a_j)\right]$ $\Delta_m=(r+\delta-2\alpha)^2-\alpha^2$ $\alpha=\frac{r+2\delta}{6}-\sqrt{\left(\frac{r+2\delta}{6}\right)^2+\frac{2}{3}}$	$\lim_{t\to\infty} x(t)=\frac{2}{\Delta_c}[\rho(r+\delta)-(a_1+a_2)]$ $\Delta_c=-\delta(r+\delta)-8$

Table 7 Individual firm investment—common and firm-specific benefits

	Markov perfect Nash equilibrium	Full coordination
Common benefits **only**	$\lim_{t\to\infty} x_1(t) = \frac{\alpha(\beta_2-\rho)-(\alpha-\delta)(\beta_1-\rho)}{(\alpha-\delta)^2-\alpha^2}$ $\lim_{t\to\infty} x_2(t) = \frac{\alpha(\beta_1-\rho)-(\alpha-\delta)(\beta_2-\rho)}{(\alpha-\delta)^2-\alpha^2}$	$\lim_{t\to\infty} x_i(t) = \frac{1}{\Delta_c}[\rho(r+\delta)-(a_1+a_2)]$ $\Delta_c = -\delta(r+\delta)-8$ That is, $\lim_{t\to\infty} x_1(t) = \lim_{t\to\infty} x_2(t)$
Common and firm-specific benefits	$\lim_{t\to\infty} x_1(t) = \frac{\alpha(\beta_2+\phi_2-\rho)-(\alpha+\theta-\delta)(\beta_1+\phi_1-\rho)}{(\alpha+\theta-\delta)^2-\alpha^2}$ $\lim_{t\to\infty} x_2(t) = \frac{\alpha(\beta_1+\phi_1-\rho)-(\alpha+\theta-\delta)(\beta_2+\phi_2-\rho)}{(\alpha+\theta-\delta)^2-\alpha^2}$ $\beta_i = \frac{2\alpha+\theta-r-\delta}{(2\alpha+\theta-r-\delta)^2-\alpha^2}W_i + \frac{-\alpha}{(2\alpha+\theta-r-\delta)^2-\alpha^2}W_j$ $\theta = \frac{(r+2\delta)\pm\sqrt{(r+2\delta)^2+8s_1}}{2}$ $\alpha = \frac{(r+2\delta-2\theta)\pm\sqrt{(r+2\delta-2\theta)^2+24}}{6}$ $W_i = \frac{\alpha(2r\rho+2\delta\rho-s_1b_1-s_2b_2)}{r+\delta-\theta} - a_i$ $\phi_1 = \frac{s_1b_1-\theta\rho}{r+\delta-\theta},\ \phi_2 = \frac{s_2b_2-\theta\rho}{r+\delta-\theta}$	$\lim_{t\to\infty} x_1(t) = \frac{4[\rho(r+\delta)-(a_1+a_2+s_2b_2)]-[\rho(r+\delta)-(a_1+a_2+s_1b_1)][\delta(r+\delta)+2s_2+4]}{[\delta(r+\delta)+2s_1+4][\delta(r+\delta)+2s_2+4]-16}$ $\lim_{t\to\infty} x_2(t) = \frac{4[\rho(r+\delta)-(a_1+a_2+s_1b_1)]-[\rho(r+\delta)-(a_1+a_2+s_2b_2)][\delta(r+\delta)+2s_1+4]}{[\delta(r+\delta)+2s_1+4][\delta(r+\delta)+2s_2+4]-16}$

Chapter 4
Dynamic Coordination for New Product Development

Abstract One of the most significant areas in supply chain coordination is new product development. In particular, it is often strategically imperative for the manufacturer and its supplier to collaborate to develop a new product successfully. Since it is becoming more and more important to take into account serviceability when developing a new product, I discuss how the firm should design an operation strategy to manufacture spare parts to honor its end-of-life warranty for its product.

Keywords Supply chain coordination · New product development · Serviceability · End-of-life warranty

Key Learning

- *How is the demand of a new product influenced by firms' innovation and price?*
  - Optimal dynamics between the competitors' technology and price levels depend on the relative magnitude of innovation and price differential coefficients.
- *How do firms compete and also cooperate together to develop a new product?*
  - We develop a dynamic model of collaboration between a manufacturer and its supplier, where the fundamental issue is, for each player, how to allocate its own resources between improving an existing product and developing a new one.
- *How should the firm determine its product price and warranty period and plan the spare parts manufacturing so as to maximize its profit and at the same time to fulfill its commitment to providing the customer with the key part continuously over the relevant decision time horizon?*

The original version of this chapter was revised. The erratum to this chapter is available at 10.1007/978-981-10-3599-9_6

B. Kim, *Optimal Control Applications for Operations Strategy*,
DOI 10.1007/978-981-10-3599-9_4

- It is not always true that the longer the EOL (end of life) warranty period, the better for the company's profitability.
- There exists an optimal EOL warranty period that balances all the relevant forces such as market demand and cost structures.

One of the most important coordination areas in supply chain management is concerned with new product development. It is not unusual to observe a case, where competitors are willing to cooperate with each other at least implicitly to grow the common market for their respective new products. It is not rare that the supply chain partners, e.g., the manufacturer and its supplier, have to make a difficult decision to allocate resources between developing a new product and improving an existing product. In this chapter, we consider these problems and try to find answers. First, we examine a case, where the demand of a new product follows an applied diffusion model influenced by innovation and price differential effects, and the potential market size expands as the technology level embodied in the product advances. For a duopoly, we set up a differential game model and derive open-loop Nash equilibrium solutions, showing that the optimal dynamics between the competitors' technology and price levels depend on the relative magnitude of innovation and price differential coefficients. If the two competitors' price differential coefficients are symmetrical, the optimal prices become constant and in general the technology levels increase. Then, we develop a dynamic model of collaboration between a manufacturer and its supplier, where the fundamental issue is, for each player, how to allocate its own resources between improving an existing product and developing a new one. We study the optimal time path of effort allocation for each player in a noncooperative setting and then look into the cooperative one. Finally, we investigate the firm's new product development strategy to determine its product price and warranty period and plan the spare parts manufacturing so as to maximize its profit and at the same time to fulfill its commitment to providing the customer with the key part continuously over the relevant decision time horizon, i.e., the product's life cycle plus its end-of-life warranty period.

1 Optimal Dynamics of Technology and Price in a Duopoly Market[1]

Technological innovation has been an important research subject in economics since Schumpeter (1947) postulated the creative response, defining the entrepreneur's function as innovation. Sengupta (2001) developed a diffusion model

[1]This section is a refined version of Kim, B., F. El Ouardighi, and S. Park (2012) 'Optimal dynamics of technology and price in a duopoly market.' *Applied Economics Letters, 19* (11), 1017–1022.

to analyze the new innovations of technology, a key element of evolutionary growth emphasized in Schumpeterian dynamics. Since new technology must be associated with new products or services, economists tried to study innovation in the context of new product development (Goel 2006), determine an optimal level of technology, that is, optimal quality level (Zhao 2000), and understand the consumer's willingness to pay for the new products or services (Chanel et al. 2006). In fact, successful new product innovation is an essential competitive advantage for a firm. But, firms often fail to launch a new product successfully. Sometimes, they introduce a new product with technology much more advanced than what the market desires (Pennings 2004; Miyazaki 2009). Sometimes, the situation is opposite: The market ignores the new product since the customers' taste for the technology is much more sophisticated. In this context, Wind and Mahajan (1997) emphasized the importance to assess the readiness of the market for the proposed new product and Bridges et al. (1995) modeled the impact of customer expectations of price and technology on the market share. Price and technology are the key variables reflecting the market readiness for the new product.

Suppose that competing firms introduce new products embodied by certain technologies, trying to determine the market's readiness to accept such products. Feichtinger and Dockner (1985) suggested a differential game model, where two firms compete for the same stock of customers: For a given level of technology, if the market readiness is higher, the price to be paid by the market would be higher than otherwise, and vice versa (Dockner and Jorgensen 1988). On the other hand, the potential market size expands as the technology (e.g., customization) level embodied in the product increases. Using the data on four generations of the IBM mainframe computer, Mahajan and Muller (1996) postulated the cumulative sales enlarged as the mainframe computer advanced in generation. Mukhopadhyay and Kouvelis (1997) proposed a theoretic differential game model to analyze the design quality and pricing decisions for a product over its product life cycle, assuming any improvements in quality would result in more industry sales.

In this section, we endeavor to explore the dynamics of technology and price for a new product in a duopoly market by taking into account those key factors discussed above.

1.1 A Differential Game Model for Duopoly

Consider a case, where two leading competitors in the market have to decide their price and technology level in a differential game context (Chintagunta and Rao 1996). We seek to obtain an open-loop Nash equilibrium solution, consistent with that the competitors decide their optimal strategies at the beginning of the game by taking into account each other's strategic choices. The open-loop Nash equilibrium solution can be compatible with a situation, where the two players are knowledgeable about the market and each other's strategic intent.

In our model, there are two competing firms involved in developing new products. The new product is an embodiment of new technology: Each firm has to decide how much technology it wants to include in its product (or how much it wants to customize the product), and the technology embodied in the product dynamically changes. We assume the two introduce their products at the same time at $t = 0$. Given the technology level embodied in the product at $t \in [0, T]$, each firm has to decide its price p_i to maximize its total cumulative profit at $t = T$.

The potential market size is determined by the two firm's technology levels $N(t) \equiv \rho_1 L_1(t) + \rho_2 L_2(t)$ simultaneously. Since the potential market size affects both firms' cumulative profits, the two players are likely motivated to collaborate, albeit in an implicit manner, so as to enlarge the total market. Then, the firms must compete against each other in order to grab a larger market share at t by dynamically adjusting their prices. Each firm's demand at t is a function of the total market size and the differential between the firm's own price and its competitor's (Shen 2006), $\dot{x}_i = ((\rho_1 L_1 + \rho_2 L_2) - (x_1 + x_2))(a_i + k_i(p_{3-i} - p_i))$: In a way, a_i represents each firm's brand power since even if the firm charges more than its competitor does, it can still attract 'some customers,' who value the firm's brand highly. The salvage value reflects not only the firm's own, but also its competitor's installed base at $t = T$, $\sigma_i x_i(T) - \delta_i x_{3-i}(T)$: Firm i's salvage value increases as its own installed base expands, but decreases as the competitor's grows.

Based on the assumptions, we develop player i's optimal control problem as follows (Table 1):

Table 1 Variables and parameters

$L_i(t)$: Technology level embodied in the new product (firm i's) $t \in [0, T]$
i: Firms $i = 1, 2$
$x_i(t)$: Cumulative sales by $t \in [0, T]$
$\dot{x}_i(t)$: Instantaneous sales at $t \in [0, T]$
$p_i(t)$: Price level at $t \in [0, T]$
c_i: Unit product cost, assuming $p_i(t) - c_i > 0$
$N \equiv \rho_1 L_1 + \rho_2 L_2$: Maximum number of adopters, the potential market size; as a linear function of L_is
ρ_i: Market size coefficient, that is, how sensitive the potential market size is to the technology level
$a_i \geq 0$: Innovation coefficient
$k_i \geq 0$: Price-related adoption (price differential) coefficient
ϕ_i: Positive constant cost related with developing the technology
σ_i: Salvage value to firm i of the unit firm i sold
δ_{3-i}: Salvage value to firm i of the unit firm 3-i (i.e., the other firm) sold

$$Max_{L_i,p_i} J_i = \int_0^T \{(p_i - c_i)((\rho_1 L_1 + \rho_2 L_2) - (x_1 + x_2))(a_i + k_i(p_{3-i} - p_i)) - \phi_i L_i^2\}\mathrm{d}t + (\sigma_i x_i(T) - \delta_i x_{3-i}(T)) \tag{1}$$

$$Subject\ to\ \ \dot{x}_i = ((\rho_1 L_1 + \rho_2 L_2) - (x_1 + x_2))(a_i + k_i(p_{3-i} - p_i)) \tag{2}$$

$$x_i(0) = x_0 \tag{3}$$

We derive open-loop Nash equilibrium solutions as follows.

Theorem 1 *An optimal solution must satisfy the following necessary conditions.*

$$p_i^* = \frac{1}{3k_i k_{3-i}}(2k_{3-i}a_i - k_i k_{3-i}\lambda_{3-i,3-i} + k_i k_{3-i}c_{3-i} - 2k_i k_{3-i}\lambda_{ii} + k_i a_{3-i} + 2k_{3-i}^2\lambda_{i,3-i} + 2k_i k_{3-i}c_i + k_i^2\lambda_{3-i,i}), \tag{4}$$

$$L_i^* = \frac{\rho_i\{(p_i - c_i + \lambda_{ii})(a_i + k_i(p_{3-i} - p_i)) + \lambda_{i,3-i}(a_{3-i} + k_{3-i}(p_i - p_{3-i}))\}}{2\phi_i}, \tag{5}$$

$$\dot{\lambda}_{ii} = \dot{\lambda}_{i,3-i} = (p_i - c_i + \lambda_{ii})(a_i + k_i(p_{3-i} - p_i)) + \lambda_{i,3-i}(a_{3-i} + k_{3-i}(p_i - p_{3-i})) \tag{6}$$

$$\lambda_{i,3-i} = \lambda_{ii} - (\sigma_i + \delta_i). \tag{7}$$

The solution also satisfies sufficient conditions.

Proof See Appendix 1.

The optimal dynamics between the competitors' technology and price levels depend on the relative magnitude of innovation and price differential coefficients. Now, we consider a simpler, yet more practical case, where the two firms' demand functions are symmetrical in price differentials, i.e., $k_1 = k_2 = \bar{k}$.

Theorem 2 *If $k_1 = k_2 = \bar{k}$, the firms' optimal prices are constant. That is,*

$$p_i^* = \frac{1}{3\bar{k}}(2a_i - \bar{k}\sigma_{3-i} + \bar{k}c_{3-i} - 2\bar{k}\sigma_i + a_{3-i} - 2\bar{k}\delta_i + 2\bar{k}c_i - \bar{k}\delta_{3-i}), \tag{8}$$

$$L_i^* = \frac{\rho_i\{(\rho_i^* - c_i + \lambda_{ii})(a_i + \bar{k}(p_{3-i}^* - p_i^*)) + \lambda_{i,3-i}(a_{3-i} + \bar{k}(p_i^* - p_{3-i}^*))\}}{2\phi_i}, \tag{9}$$

$$\dot{\lambda}_{ii} = \dot{\lambda}_{i,3-i} = (p_i^* - c_i + \lambda_{ii})(a_i + \bar{k}(p_{3-i}^* - p_i^*)) + \lambda_{i,3-i}(a_{3-i} + \bar{k}(p_i^* - p_{3-i}^*)) \tag{10}$$

$$\lambda_{i,3-i} = \lambda_{ii} - (\sigma_i + \delta_i). \tag{11}$$

Proof See Appendix 1.

Theorem 2 implies that when there is no difference between the price differential coefficients, it is useless for each company to try to outsmart its competitor by adjusting its price: They should compete with decision variables other than the prices. As a result, the two companies stick with fixed prices and the price differential effects remain steady throughout the decision horizon—see Table 2 and Fig. 1, where company 1 can afford to charge a higher price since its innovation coefficient is much larger, which also explains why company 1 increases its technology in a much more accelerating manner. It is intriguing to observe that the companies improve their technologies always (see Theorem 3).

Theorem 3 *Given* $k_1 = k_2 = \bar{k}, \dot{L}_i^* = \frac{dL_i^*}{d_t} \geq 0$ *if it holds that*

$$\lambda_{ii} \geq \frac{(\sigma_i + \delta_i)\left(a_{3-i} + \bar{k}(p_i^* - p_{3-i}^*)\right) - (p_i - c_i)(a_i + \bar{k}(p_{3-i}^* - p_i^*))}{a_i + a_{3-i}}. \tag{12}$$

Proof From (9), $\dot{L}_i^* = \frac{\rho_i\{\lambda_{ii}(a_i + \bar{k}(p_{3-i}^* - p_i^*)) + \dot{\lambda}_{i,3-i}(a_{3-i} + \bar{k}(p_i^* - p_{3-i}^*))\}}{2\phi_i} = \frac{\rho_i(a_i + a_{3-i})}{2\phi_i}\dot{\lambda}_{ii}$ due to (10). Therefore, $\dot{L}_i^* \geq 0$ if $\dot{\lambda}_{ii} \geq 0$. Now, using (10) and (11), we can show $\dot{\lambda}_{ii} \geq 0$ if and only if $\lambda_{ii} \geq \frac{(\sigma_i + \delta_i)\left(a_{3-i} + \bar{k}(p_i^* - p_{3-i}^*)\right) - (p_i - c_i)\left(a_i + \bar{k}(p_{3-i}^* - p_i^*)\right)}{a_i + a_{3-i}}$.

Table 2 Parameter values

Player	σ	a	k	ρ	ϕ	c	δ
1	20	0.02	0.00030	200	10	200	20
2	19	0.01	0.00030	190	9	190	19

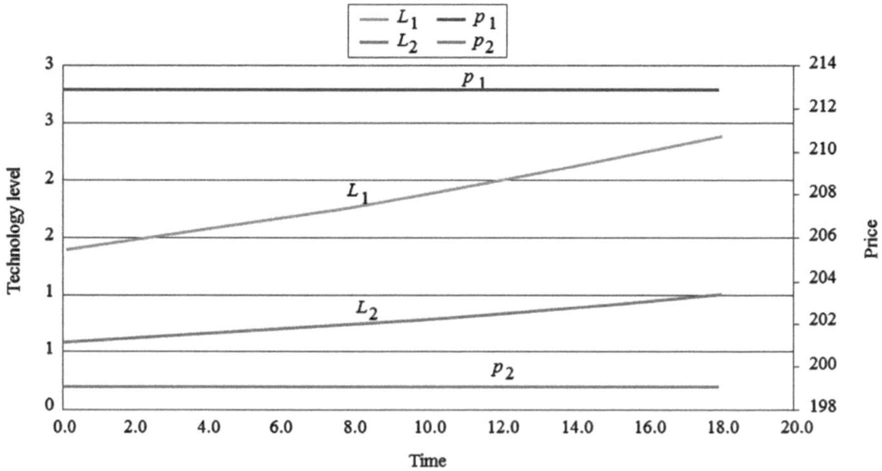

Fig. 1 Dynamic changes of L^* and p^*

Under a normal condition, we expect $\dot{L}_i^* \geq 0$, i.e., the firm always improves its technology: Since λ_{ii} is the marginal valuation of x_i from company i's perspective, it is always better to have one more unit of x_i, and thus, $\lambda_{ii} \geq 0$. Assuming a_i and a_{3-i} are not very different, (12) tends to hold as long as the unit production cost c_i is not very large compared with p_i. We have conducted numerical analyses using a wide range of realistic and reasonable parameter values and found out that $\frac{dL_i^*}{\mathrm{d}t} \geq 0$ always for the duopoly.

1.2 *Managerial Implications and Conclusions*

In this section, we have considered two important questions for successful new product strategy, i.e., *the technology* (e.g., *customization*) *level to be embodied in the product* and *the price of the product to be charged.* For the duopoly case, with a differential game model, we specified optimal dynamic paths of key decision variables and presented a numerical example. When the price differentials in the two competitors' demand functions have a symmetrical effect, the duopolistic companies seem to improve their technologies continuously while keeping their prices constant: The competitors share a common interest to enlarge the potential market size in the first place before getting to the point of how to divide the pie.

2 Supplier–Manufacturer Collaboration on New Product Development[2]

Collaboration between supplier and manufacturer is prevalent and can be of diverse forms. A supplier often collaborates with its customer, e.g., a manufacturing company, on developing new products. But there is a potential trade-off in their endeavor. On the one hand, there needs to be a short-term collaboration between the supplier and the manufacturer on the manufacturer's current product. For instance, in order to gain a larger market share for the existing product, it would be important for the product quality to improve or for the price to decrease. To do so, the manufacturer expects its supplier to support its efforts to improve the existing product value. On the other hand, the manufacturer needs its supplier's help in developing a new product, which is necessary for the manufacturing company to continue to grow in the market (Lau et al. 2010; Peng et al. 2014).

Compared with the joint efforts to improve the existing product's quality, collaborating for new product development is a long-term endeavor. For example,

[2]This section is based on part of Kim, B. and F. El Ouardighi (2007) 'Supplier–manufacturer collaboration on new product development.' *Annals of Dynamic Games, 9*, 527–547.

consider the case of SKT, a Korean telecommunications company. In the early 1990s, the company had struggled to make a transition from an analog to a digital mobile phone system. While using the analog technology, SKT tried to improve the telecommunications quality with the help from its key suppliers. At the same time, since SKT knew the new digital technology would replace the old one, it started to invest in developing the digital technology and asked those suppliers (who were also involved in improving quality of the analog telecommunications system) to participate in the project. Then, the relevant decision problem was how to allocate resources between the current technology (product quality) improvement and the development of new technology (product). This was the problem faced by not only SKT, but also its supplier.

Now, the fundamental issue facing the supplier and the manufacturer, respectively, is how to allocate resources between improving the existing product and developing a new one. By improving the existing product, they expect to earn more *current profit*, i.e., their short-term profitability will increase. By enhancing the chance to have a successful new product development, they can expect to earn *future profit*, which is vital for their continuous growth. Since they have to allocate their *present* resources between *current* operations and *future* possibility, they are facing a decision problem of trade-off.

2.1 Model Formulation

We consider a supply chain with two firms, a manufacturer and its supplier, both involved in two major activities, i.e., quality improvement of an existing product and development of a new product.

We first describe the dynamics of the quality improvement activity. Let $q(t) > 0$ denote the quality level at time t of the product on the market (i.e., current product). Quality is a state variable incremented by the respective efforts of the manufacturer, $u_1(t)$, and the supplier, $u_2(t)$, that is, $\dot{q}(t) = g(u_1(t), u_2(t)), q(0) = q_0 > 0$.

For the dynamics of new product development (NPD), let $z(t) \geq 0$ be the cumulative level of knowledge at time t on a product in development (i.e., future product). The NPD activity is also a state variable incremented by the respective contributions of both players, $v_i(t), i = 1, 2,$ that is, $\dot{z}(t) = h(v_1(t), v_2(t)), z(0) = z_0 \geq 0$.

Under the negative exponential distribution, the breakthrough cumulative probability at t is given by $\phi(z(t)) = 1 - e^{\lambda z(t)}$, which indeed depends on the cumulative effort z. Moreover, $\frac{d\phi(z(t))/\mathrm{d}t}{(1-\phi(z))} = \lambda \dot{z}$ is the conditional probability of completion, which is a standard setting as we are dealing with instantaneous probabilities (Kamien and Schwartz 1978; Reinganum 1984).

Using a finite planning horizon, i.e., $t \in [0, T]$, the objective criterion of the manufacturer is

$$Max_{p_1,u_1,v_1} J^1 = \int_0^T \{[(p_1(t) - p_2)f(p_1(t), q(t)) - c_1(u_1(t)) - d_1(v_1(t))] \times (1 - \phi(z(t))) + \omega\phi_z\dot{z}(t)N\}dt + \theta_1 z(T).$$

For simplicity, we assume a short enough planning horizon so that the discounting effect can be disregarded. At time *t*, provided the new product is not completed yet, the manufacturer's net profit from the existing product is $(p_1 - p_2)f(p_1, q) - c_1(u_1) - d_1(v_1)$ with probability $(1 - \phi(z))$. It is composed of the following elements: gross profit margin $(p_1 - p_2)$, that is, sales price minus (constant) transfer price to the supplier, times sales of the existing product, $f(p_1, q)$, minus expense to improve the existing product, $c_1(u_1)$, minus expense to develop the new product, $d_1(v_1)$. The assumption of a constant transfer price follows from the fact that when manufacturers and suppliers decide to collaborate, the transfer price to the supplier is one of the terms that are usually fixed at the time of signing the contract, especially when the contract period is short term. We assume that the demand function $f(p_1, q)$ is influenced negatively by price, i.e., $f_{p_1} < 0$, and positively by quality, i.e., $f_q > 0$.

In addition, with probability $d\phi(z(t)) = \phi_z\dot{z}(t)dt = \phi_z h(v_1(t), v_2(t))dt$, the new product will become available during the time interval $(t, t + dt)$, providing a future utility stream with discounted value ωN at time *t*. The constant $\omega, 0 \leq \omega \leq 1$ (respectively,$0 \leq 1 - \omega \leq 1$) denotes the constant rate which should be allocated to the manufacturer (respectively, the supplier). On the other hand, *N* represents the total future benefits from the new product, and from the (unknown) time, it becomes available forward. We assume that the players have a priori knowledge about the size of the lump-sum benefit from the new product development success, so that *N* may be considered as constant. Note that the manufacturer and the retailer agree beforehand on each one's share in the revenues of the new product. This refers to the fact that partnering companies generally want to design their collaboration contract in a predictable manner as much as possible. Predetermining the profit sharing scheme is part of such an effort. For our analysis, we first define ω *ex ante* and analyze our model. Finally, letting $\theta_1 \geq 0$, $\theta_1 z(T)$ represents the salvaged value of the stock of knowledge at the end of the planning horizon.

Using similar arguments, the supplier's problem is

$$Max_{u_2,v_2} J^2 = \int_0^T \{[p_2 f(p_1(t), q(t)) - c_2(u_2(t)) - d_2(v_2(t))] \times (1 - \phi(z(t))) + (1 - \omega)\phi_z\dot{z}(t)N\}dt + \theta_2 z(T).$$

Then, the manufacturer controls its sales price and its quality improvement and knowledge accumulation efforts, while the supplier controls its quality improvement and knowledge accumulation efforts.

For the demand function, the following multiplicative expression is introduced:

$f(p_1(t), q(t)) = eq(t)(a - bp_1(t)),\quad a, b, e > 0,$ where demand is linearly decreasing with price and increasing with quality.

On the other hand, for the probability distribution function, we assume that the probability of successful development by a given date is known to both players and should be given by a negative exponential distribution, that is, $\phi(z(t)) = 1 - e^{-\lambda z(t)}, \lambda > 0$ and $\frac{d\phi(z(t))}{dt} = \lambda\dot{z}(t)e^{-\lambda z(t)}$. Thus, the conditional probability that the new product is completed immediately beyond time t given no successful development until time t, i.e., the hazard rate, is $\frac{\lambda\dot{z}(t)e^{-\lambda z(t)}}{1-\phi(z(t))} = \lambda\dot{z}(t)$.

In addition, for tractability, the following linear additive specifications are, respectively, introduced: $g(u_1(t), u_2(t)) = \sum_{i=1}^{2} \alpha_i u_i(t)$, $h(v_1(t), v_2(t)) = \sum_{i=1}^{2} \beta_i v_i(t)$, where $\alpha_i > 0, i = 1, 2$ is the marginal impact of player i's effort to the quality improvement process, and $\beta_i > 0, i = 1, 2$ is the marginal influence of player i's contribution to the development process.

Finally, we assume convex increasing cost functions for both players, with explicit functions, respectively, for quality improvement effort, $c_i(u_i(t)) = \frac{c_i(u_i(t))^2}{2}$, $c_i > 0, i = 1, 2,$ and for new product development effort, $d_i(v_i(t)) = \frac{d_i(v_i(t))^2}{2}$, $d_i > 0, i = 1, 2.$

Hence, the manufacturer's problem is written as

$$Max \int_0^T \{[(p_1(t) - p_2)(a - bp_1(t))eq(t) - \frac{c_1(u_1(t))^2}{2} - \frac{d_1(v_1(t))^2}{2}] + \lambda\omega\dot{z}(t)N\}$$
$$\times\, e^{-\lambda z(t)}dt + \theta_1 z(T)$$

$$Subject\ to\ \ \dot{q}(t) = \sum_{i=1}^{2} \alpha_i u_i(t), q(0) = q_0 > 0,$$
$$\dot{z}(t) = \sum_{i=1}^{2} \beta_i v_i(t), z(0) = z_0 \geq 0.$$

To analyze the game, it is convenient to introduce the state transformation $y(t) = e^{-\lambda z(t)}$, where $y(t) \in [0, 1]$.

Differentiating both sides of this equation with respect to time, we get

$$\dot{y}(t) = -\lambda y(t)\dot{z}(t) = -\lambda y(t)\sum_{i=1}^{2} \beta_i v_i(t).$$

The manufacturer's problem can now be rewritten in terms of the new state variable:

$$Max_{p_1,u_1,v_1} J^1 = \int_0^T \left\{ \left[(p_1(t) - p_2)(a - bp_1(t))eq(t) - \frac{c_1(u_1(t))^2}{2} - \frac{d_1(v_1(t))^2}{2} \right] \times y(t) - \omega\dot{y}(t)N \right\} dt - \frac{\theta_1}{\lambda} lny(T),$$

$$Subject\ to\ \ \dot{q}(t) = \sum\nolimits_{i=1}^{2} \alpha_i u_i(t), q(0) = q_0 > 0,$$
$$\dot{y}(t) = -\lambda y(t) \sum\nolimits_{i=1}^{2} \beta_i v_i(t), y(0) = y_0 \in [0, 1).$$

2.2 Noncooperative Game

The game defined above combines linear state and exponential properties (Dockner et al. 2000), which results in an interaction between the state variables. This particular structure precludes any useful characterization of closed-loop equilibrium strategies. In this sense, we confine our interest to an open-loop Nash equilibrium, which offers more tractable and flexible implications.

Skipping the time index for simplicity, the Hamiltonian of the manufacturer is

$$H^1 = \left\{ (p_1 - p_2)(a - bp_1)eq - \frac{c_1 u_1^2}{2} - \frac{d_1 v_1^2}{2} + \lambda\left[\omega N - \mu_2^1\right] \sum\nolimits_{i=1}^{2} \beta_i v_i \right\} y + \mu_1^1 \sum\nolimits_{i=1}^{2} \alpha_i u_i, \quad (13)$$

where $\mu_j^1(t)$ are the costate variables, $j = 1, 2$.

Assuming interior solutions, necessary conditions for optimality include

$$H_{p_1}^1 = (a - 2bp_1 + bp_2)eqy = 0, \quad (14)$$

$$H_{u_1}^1 = -c_1 u_1 y + \mu_1^1 \alpha_1 = 0, \quad (15)$$

$$H_{v_1}^1 = \left\{ \lambda\beta_1 \left[\omega N - \mu_2^1\right] - d_1 v_1 \right\} y = 0. \quad (16)$$

Lemma 1 *An optimal noncooperative price satisfies*

$$p_1 = \frac{1}{2}\left[p_2 + \frac{a}{b}\right]. \quad (17)$$

Proof The proof is obvious from (14).

This is a myopic rule that is inversely related to the marginal influence of transfer price on demand. The sales price increases with transfer price to the supplier, and this rule amounts to pricing from the transfer price to the supplier.

Substituting (17) for optimal sales price into (13), the costate variables have the following respective dynamics:

$$\dot{\mu}_1^1 = \frac{e}{-4b}(a - bp_2)^2 y, \tag{18}$$

$$\dot{\mu}_2^1 = -\left[\frac{e}{4b}(a - bp_2)^2 q - \frac{c_1 u_1^2}{2} - \frac{d_1 v_1^2}{2}\right] - \lambda\left[\omega N - \mu_2^1\right]\sum_{i=1}^{2} \beta_i v_i, \tag{19}$$

$\mu_1^1(T) = 0$ and $\mu_2^1(T) = -\frac{\theta_1}{\lambda y(T)}$ being the transversality conditions.

In (18), $\dot{\mu}_1^1 \leq 0$ and $\mu_1^1(T) = 0$ both imply $\mu_1^1(t) \geq 0 | t \leq T$, which indicates that the improvement activity has a positive influence on the manufacturer's profit. Note that $y(t) \mapsto 1$ if $z(t) \mapsto +\infty$, which is unlikely to be verified over a (short) finite time horizon. Solving for $u_1(t)$ in (15) should then result in $u_1(T) = 0$, that is, $u_1(t)$ should be decreasing over time.

In (19), for a nonnegative instantaneous profit, $\dot{\mu}_2^1|_{\mu_2^1 \leq 0} < 0$. Given the transversality condition $\mu_2^1(T) = -\frac{\theta_1}{\lambda y(T)}$, the sign of $\mu_2^1(t)$ is ambiguous for $\theta_1 > 0$. However, for $\theta_1 = 0$, $\mu_2^1(T) = 0$, which results in $\mu_2^1(T) \geq 0 | t \leq T$. Solving for $v_1(t)$ in (16) yields $v_1(t) = \frac{\lambda \beta_1}{d_1}\left[\omega N + \frac{\theta_1}{\lambda y(T)}\right] > 0$, $\theta_1 \geq 0$, which shows that $v_1(t)$ should be increasing over time until it reaches its maximum value at $t = T$.

We now turn to the supplier's problem. Using (17) for sales price, the supplier's Hamiltonian is

$$H^2 = \left\{\frac{p_2}{2}(a - bp_2)eq - \frac{c_2 u_2^2}{2} - \frac{d_2 v_2^2}{2} + \lambda\left[(1 - \omega)N - \mu_2^2\right] \times \sum_{i=1}^{2} \beta_i v_i\right\} y + \mu_1^2 \sum_{i=1}^{2} \alpha_i u_i, \tag{20}$$

where $\mu_j^2(t)$ are the costate variables, $j = (1, 2)$.

Assuming interior solutions, necessary conditions for optimality include

$$H_{u_2}^2 = -c_2 u_2 y + \mu_1^2 \alpha_2 = 0, \tag{21}$$

$$H_{v_2}^2 = \left\{\lambda \beta_2\left[(1 - \omega)N - \mu_2^2\right] - d_2 v_2\right\} y = 0. \tag{22}$$

The costate variables have the following respective dynamics:

$$\dot{\mu}_1^2 = -\frac{ep_2}{2}(a - bp_2)y, \tag{23}$$

$$\begin{aligned}\dot{\mu}_2^2 = &-\left[\frac{p_2}{2}(a - bp_2)eq - \frac{c_2 u_2^2}{2} - \frac{d_2 v_2^2}{2}\right] \\ &- \lambda\left[(1-\omega)N - \mu_2^2\right]\sum\nolimits_{i=1}^{2}\beta_i v_i,\end{aligned} \tag{24}$$

$\mu_1^2(T) = 0$ and $\mu_2^2(T) = -\frac{\theta_2}{\lambda y(T)}$ being the transversality conditions.

In (23), $\dot{\mu}_1^2 \leq 0$ and $\mu_1^2(T) = 0$ both provide $\mu_1^2(t) \geq 0$ whatever $t \in [0, T]$, which means that the improvement process has a positive marginal effect on the supplier's profit. As for the manufacturer, solving for $u_2(t)$ in (21) should result in $u_2(T) = 0$, i.e., $u_2(t)$ should also be decreasing over time.

In (24), given a nonnegative instantaneous profit, we obtain $\dot{\mu}_2^2|_{\mu_2^2 \leq 0} < 0$. Given the transversality condition $\mu_2^2(T) = -\frac{\theta_2}{\lambda y(T)}$, the sign of $\mu_2^2(t)$ is unclear. However, for $\theta_2 = 0, \mu_2^2(T) = 0$, which results in $\mu_2^2(t) \geq 0 | t \in [0, T]$. Solving for $v_2(t)$ in (22) yields $v_2(T) = \frac{\lambda\beta_2}{d_2}\left[(1-\omega)N - \frac{\theta_2}{\lambda y(T)}\right] > 0$, $\theta_2 \geq 0$, which shows that $v_2(t)$ should be increasing over time until maximum value is reached at $t = T$.

Proposition 1 *Qualitative solutions for the noncooperative improvement effort of the manufacturer and the supplier are respectively,*

$$\dot{u}_1 = \lambda u_1 \sum\nolimits_{i=1}^{2}\beta_i v_i - \frac{e\alpha_1}{4bc_1}(a - bp_2)^2, \quad u_1(T) = 0, \tag{25}$$

$$\dot{u}_2 = \lambda u_2 \sum\nolimits_{i=1}^{2}\beta_i v_i - \frac{e\alpha_2 p_2}{2c_2}(a - bp_2), \quad u_2(T) = 0, \tag{26}$$

Proof Equations (15) and (21) are differentiated totally with respect to time to yield, after elementary manipulations, (25) and (26).

In (25) and (26), as $\dot{u}_i|_{u_i=0} < 0$ and $u_i(T) = 0$, $i = 1, 2$, the optimal improvement effort of both players is decreasing over time. The players make an effort to improve the current product quality more during the early period than in the later period since, as time passes, it becomes more likely to successfully complete the new product development. The decline of the players' improvement effort is due to the negative influence of marginal demand with respect to quality. However, the rate of change of players' improvement effort is positively affected by the speed of NPD. In other words, the innovation effort by both players slows down the decrease of each player's improvement effort. Given $u_i(T) = 0$, a negative influence of both players' innovation effort on each player's improvement effort is obtained (Fig. 2).

In the same way, a marginal increase of the supplier's transfer price reduces the manufacturer's improvement effort, i.e., $\dot{u}_{1p_2} = \frac{e\alpha_1}{2c_1}(a - bp_2) > 0$. Conversely, for

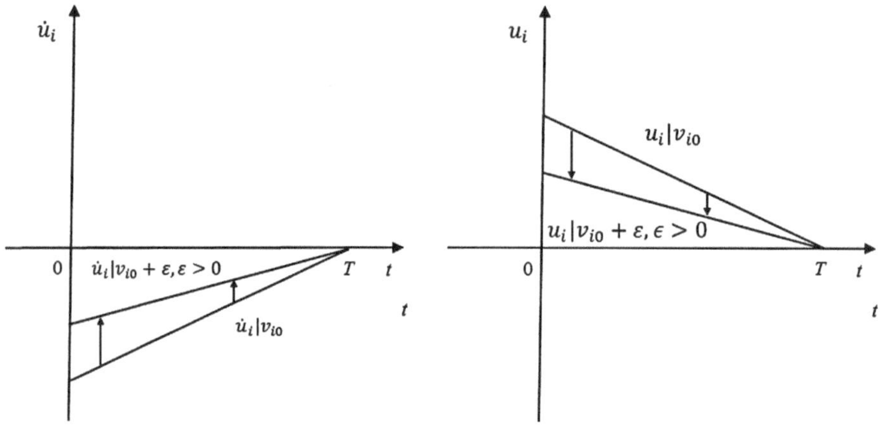

Fig. 2 Marginal impact of player's innovation effort on each one's own improvement effort

sufficiently small values of b, the supplier's transfer price has a positive impact on its own improvement effort, i.e., $\dot{u}_{2p_2}=-\frac{e\alpha_2}{2c_2}(a-2bp_2)<0$. This result is expected since there is a tension between the manufacturer and the supplier regarding the transfer price, i.e., the higher the transfer price, the smaller the manufacturer's profit and the larger the supplier's profit, and vice versa.

Proposition 2 *Qualitative solutions for the noncooperative innovation effort of the manufacturer and the supplier are respectively*

$$\dot{v}_1=\lambda\left\{v_1\left[\frac{\beta_1 v_1}{2}+\beta_2 v_2\right]+\frac{\beta_1}{d_1}\left[\frac{1}{4b}(a-bp_2)^2 eq-\frac{c_1u_1^2}{2}\right]\right\},$$
$$v_1(T)=\frac{\lambda\beta_1}{d_1}\left[\omega N+\frac{\theta_1}{\lambda y(T)}\right]>0, \tag{27}$$

$$\dot{v}_2=\lambda\left\{v_2\left[\beta_1 v_1+\frac{\beta_2 v_2}{2}\right]+\frac{\beta_2}{d_2}\left[\frac{p_2}{2}(a-bp_2)eq-\frac{c_2u_2^2}{2}\right]\right\},$$
$$v_2(T)=\frac{\lambda\beta_2}{d_2}\left[(1-\omega)N+\frac{\theta_2}{\lambda y(T)}\right]>0, \tag{28}$$

Proof Equations (16) and (22) are differentiated totally with respect to time to yield, after elementary manipulations, (27) and (28).

As expected, the larger the sharing ratio of the new product profit, ω, the higher (lower) the manufacturer's (supplier's) profitability from the new product. In (27) and (28), each player's innovation effort increases over time with the speed of knowledge accumulation. We can provide two intuitive explanations for this result. First, as time passes, the probability to have the innovation increases, and thus, the dollar spent on the innovation becomes more valuable than that on the existing product's quality improvement. Second, the benefit the firm can get from selling the existing product is limited and will diminish as time passes. Thus, as time progresses, the marginal value of the existing product decreases, whereas that of the innovation increases.

In addition, the speed of each player's innovation effort depends positively on its own instantaneous profit from the current product. In this sense, the higher the each player's 'current' profit, the faster its own search for 'future' profit. In other words, as time passes, it becomes more likely for NPD to be accomplished. As for quality improvement efforts, the speed of innovation effort by the manufacturer (supplier) is negatively (positively) affected by the supplier's transfer price.

2.3 Cooperative Game

We now look at the cooperative setting in which the manufacturer and its supplier intend to coordinate the two activities (quality improvement and innovation process) as an essential part of a supply chain. The combination of individual players' objective criteria provides the following optimal control problem:

$$Max_{p_1,u_1,u_2,v_1,v_2}\, J = \int_0^T \left\{\left[p_1(t)(a-bp_1(t))eq(t) - \sum_{i=1}^2 \frac{c_i(u_i(t))^2}{2} - \sum_{i=1}^2 \frac{d_i(v_i(t))^2}{2}\right]y(t) - \dot{y}(t)N\right\}dt - \frac{\theta}{\lambda} lny(T)$$

Subject to $\dot{q}(t) = \sum_{i=1}^2 \alpha_i u_i(t), q(0) = q_0 > 0,\ \dot{y}(t) = -\lambda y(t)\sum_{i=1}^2 \beta_i v_i(t),\ y(0) = y_0 \in [0,1)$, where $\theta = \sum_{i=1}^2 \theta_i$.

Skipping the time index for simplicity, the Hamiltonian associated to the coordinated chain is

$$H = \left\{p_1(a-bp_1)eq - \sum_{i=1}^2 \frac{c_i u_i^2}{2} - \sum_{i=1}^2 \frac{d_i v_i^2}{2} + \lambda[N-\mu_2]\sum_{i=1}^2 \beta_i v_i\right\} \times y + \mu_1 \sum_{i=1}^2 \alpha_i u_i, \tag{29}$$

where μ_j are the costate variables, $j = 1, 2$.

Assuming interior solutions, necessary conditions for optimality include

$$H_{p_1} = [a - 2bp_1]eqy = 0, \tag{30}$$

$$H_{u_i} = -c_i u_i y + \mu_1 \alpha_i = 0, \tag{31}$$

$$H_{v_i} = \{\lambda\beta_i[N - \mu_2] - d_i v_i\}y = 0. \tag{32}$$

Lemma 2 *The optimal cooperative price is lower than the optimal noncooperative price.*

Proof Solving for p_1 in (30) and comparing with (17) yield

$$p_1^{\Theta} = \frac{a}{2b} < \frac{1}{2}\left[p_2 + \frac{a}{b}\right], \tag{33}$$

where the superscript Θ is for the cooperative solution.

This result states that the coordination of a supply chain results in a higher consumer's surplus, i.e., a larger demand for the current product, than in a decentralized decision-making system. This reflects a mitigation of the double marginalization practice which is typical in uncoordinated chains (Durham 2000).

Lemma 3 *Player i's relative optimal quality improvement effort is*

$$\frac{u_i}{u_{-i}} = \frac{c_{-i}\alpha_i}{c_i\alpha_{-i}}, \tag{34}$$

where the subscript–i is for the other player.

Proof The proof is obvious from (31).

In (34), player i's relative effort has a constant value that depends on the ratio of player i's efficiency (α_i/c_i) to the other player's efficiency (α_{-i}/c_{-i}) in improving the current product quality. What this rule amounts to is that the more efficient player contributes more to the quality improvement activity.

Lemma 4 *Player i's relative optimal innovation effort is*

$$\frac{v_i}{v_{-i}} = \frac{d_{-i}\beta_i}{d_i\beta_{-i}}, \tag{35}$$

where the subscript–i is for the other player.

Proof The proof is obvious from (32).

Here also, player i's relative effort is constant and depends on the ratio of its own efficiency (β_i/d_i) to the other player's efficiency (β_{-i}/d_{-i}) in NPD. In asymmetric conditions, the more efficient player contributes more to the innovation activity.

Substituting (33) for optimal price, the costate variables have the following dynamics:

$$\dot{\mu}_1 = -\frac{a^2 ey}{4b}, \tag{36}$$

$$\dot{\mu}_2 = -\left[\frac{a^2 eq}{4b} - \sum_{i=1}^{2} \frac{c_i u_i^2}{2} - \sum_{i=1}^{2} \frac{d_i v_i^2}{2}\right] - \lambda[N - \mu_2] \sum_{i=1}^{2} \beta_i v_i, \tag{37}$$

$\mu_1(T) = 0$ and $\mu_2(T) = -\frac{\theta}{\lambda y(T)}$ being the transversality conditions.

In (36), $\dot{\mu}_1 \leq 0$ and $\mu_1(T) = 0$ imply $\mu_1(t) \geq 0 | t \leq T$. As for the noncooperative setting, $u_i^{\Theta}(T) = 0$ should result from (31), that is, the cooperative improvement effort of player *i*, $u_i^{\Theta}(t)$, should be decreasing over time.

In (37), for a nonnegative instantaneous profit, $\dot{\mu}_2|_{\mu_2 \leq 0} < 0$. If $\theta = 0$, the transversality condition is $\mu_2(T) = 0$, which results in $\mu_2(t) \geq 0 | t \leq T$. From (32), we have $v_i^{\Theta}(T) = \frac{\lambda \beta_i}{d_i}\left(N + \frac{\theta}{\lambda y(T)}\right) > 0$, which suggests that v_i^{Θ} is increasing over time until it reaches its maximum value at $t = T$.

For similar abilities of the two partners in the NPD activity, the respective innovation efforts are equal at the end of the planning horizon. Furthermore, *all things being equal*, the terminal innovation effort is larger in the cooperative than in the noncooperative setting.

Proposition 3 *Qualitative solutions for player i's cooperative improvement and innovation efforts are respectively*

$$\begin{aligned} \dot{u}_i^{\Theta} &= \lambda u_i \sum_{i=1}^{2} \beta_i v_i - \frac{a^2 e \alpha_i}{4bc_i}. \\ u_i^{\Theta}(T) &= 0, \end{aligned} \tag{38}$$

$$\begin{aligned} \dot{v}_i^{\Theta} &= \lambda \left\{ v_i \left[\frac{\beta_i v_i}{2} + \beta_{-i} v_{-i}\right] + \frac{\beta_i}{d_i}\left[\frac{a^2 eq}{4b} - \sum_{i=1}^{2} \frac{c_i u_i^2}{2} - \frac{d_{-i} v_{-i}^2}{2}\right]\right\}, \\ v_i^{\Theta}(T) &= \frac{\lambda \beta_i}{d_i}\left(N + \frac{\theta}{\lambda y(T)}\right). \end{aligned} \tag{39}$$

Proof Equations (31) and (32) are differentiated totally with respect to time to yield, after elementary manipulations, (38) and (39).

In (38), as $\dot{u}_i^{\Theta}|_{u_i = 0} < 0$ and $u_i^{\Theta}(T) = 0, i = (1, 2)$, the cooperative improvement effort of both players is decreasing over time. The decline of the players' improvement effort results from the negative influence of marginal demand with respect to quality. As marginal demand with respect to quality is larger, the players' improvement effort decreases faster in the cooperative setting than in the noncooperative setting. All things being equal, this implies a larger initial value of improvement effort in the cooperative case. As for the noncooperative scenario, the speed of knowledge accumulation has a negative influence on each player's improvement effort.

2.4 Conclusion

The issue in this section was how the collaborating firms in a supply chain (i.e., a manufacturer and its supplier) allocate resources between improving a marketed product and developing a new product. In this respect, a differential game model for both noncooperative and cooperative settings was designed. The following managerial implications emerged.

- Whatever the nature of equilibrium (noncooperative or cooperative), the optimal strategy implies that both players make initially high and decreasing efforts in quality improvement and initially low and increasing efforts in NPD. The rationale for such behaviors is to achieve maximum profits from the current product at the beginning of the game so as to invest them increasingly in the future product as time goes by.
- In a coordinated chain, the allocation of efforts is determined by the rule according to which, for each activity, the more efficient player contributes more.
- In an uncoordinated chain, the allocation of efforts by each player results from the comparison of its own implicit benefit in one activity with that in the other activity, rather than from the comparison between its own efficiency with that of the other player in the same activity. From the supplier's perspective, the higher the transfer price, the heavier the effort in quality improvement. For the manufacturer, the larger the part of profits from the new product, the heavier the effort in NPD.
- In most configurations, the uncoordinated chain pays more attention to innovation, while the coordinated chain favors the quality of the existing product.

3 New Product and Warranty Strategy[3]

As the competition intensifies in the global market, the product life cycle gets shortened fast and firms try to introduce new products more frequently as one of their key strategic responses. In addition, as the customers become more concerned about the product's entire life cycle rather than only the time when they purchase the product, e.g., whether and how to replace and/or recycle failed parts of the product throughout its total life, firms are offering more extensive warranty to customers in order to increase sales (Chen et al. 2012; Wei et al. 2015). For instance, by offering an extensive warranty for its cars, Hyundai Motor was able to change consumers' perception so that the market started to believe the company's quality has improved to match the very best in the industry (Business Week 2004).

---

[3]This section is a refined version of Kim, B. and S. Park (2008) 'Optimal pricing, EOL (end of life) warranty, and spare parts manufacturing strategy amid product transition.' *European Journal of Operational Research, 188* (3), 723–745.

Consider a cellular phone manufacturer. This company manufactures and sells a series of cellular phones, each of which has a relatively short life cycle. The company considers developing a new model, NM-1. It estimates NM-1's life cycle will be T. That is, after T, the company will stop selling the model. NM-1 consists of multiple parts, one of which is the company's proprietary microprocessor, mp-1: All the other parts except for mp-1 can be found in the market, if spare parts are needed. That is, the company must produce mp-1 and guarantee providing it to customers for a certain period of time, i.e., the warranty period, τ: During the warranty period, the company must have mp-1 in stock so as to meet the demand for the microprocessor. Therefore, when a customer considers buying NM-1, she probably takes into account the length of the warranty period, during which she can get her cellular phone repaired when the key microprocessor breaks down. In effect, the warranty length can affect the customer's buying decision.

Consider a realistic context, when $T > \tau$ and a customer bought the product at t_u so that $t_u < T - \tau$. If the company services the product only for τ period after the customer bought the product, i.e., $t_u \leq t \leq t_u + \tau$, it might happen that the product cannot be serviced if it breaks down at t_r such that $t_u + \tau < t_r < T$, even though the company still sells the product until T. From the customer's point of view, this policy does not look appealing. Therefore, it is more reasonable to assume that if the product's life cycle, T, is relatively short, the warranty period, τ, should mean that the company will service the customer with the key part until $T + \tau$. That is, the company guarantees availability of the key part for τ after it discontinues producing the current product at T. In effect, the company guarantees after-sales services without time limitation as long as the product is being sold in the market and also offers 'EOL (end-of-life)' services for τ period after the current product is discontinued. This kind of arrangement is the 'EOL warranty': It seems realistic and reasonable, in particular when the product's life cycle is relatively short.

Now, the company has to make decision on several variables. What should be the optimal warranty period? How much should the company charge for the product? How many spare parts should the company produce during the product's life cycle and also during the EOL warranty period, if any? These decisions are interrelated with each other, and the company has to take into account many factors when trying to find an optimal solution.

3.1 The Two-Stage Optimal Control Theory Model

In order to investigate the dynamic relationships among optimal EOL warranty periods, optimal pricing, and optimal manufacturing strategy for spare parts, we employ a two-stage optimal control theory model. This model has been widely used when a decision problem is involved with two separate decision-making periods. Most notably, Tomiyama (1984) provided a theoretical foundation for the two-stage optimal control theory model by studying a two-stage optimal control problem characterized by the change in the state equation and the performance index at an

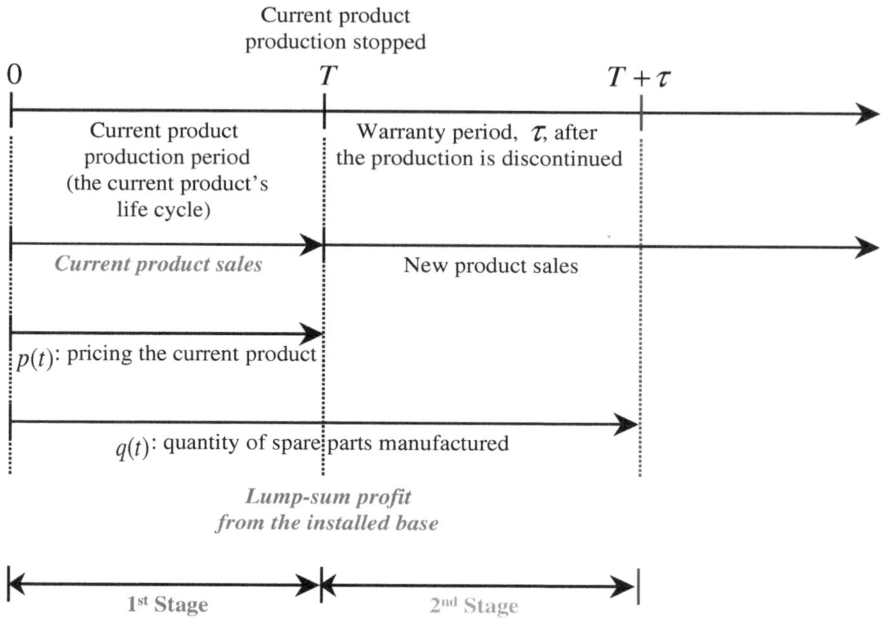

Fig. 3 Decision problem—key control variables

unspecified switching time. Afterward, Rossana (1985) postulated necessary conditions for the solution of two-stage optimal control problems, where the switching point is a choice variable.

Rather than discussing the two-stage optimal control theory model[4] separately, we would like to explain the methodology while developing and solving the actual problem in this section. Our decision problem is depicted in Fig. 3. The company makes and sells the current product until T (the 1st stage), but guarantees after-sales services for a key part of the product for τ after the product is discontinued at T (the 2nd stage). Therefore, the key decision variable for the first stage is the pricing, $p(t)$, while the firm has to decide how many units of spare parts to make throughout the entire decision horizon, i.e., $T + \tau$. We will further assume that the product life cycle T itself is a decision variable the firm has to decide. Then, the firm needs to determine both optimal T and τ at the same time.

Figure 4 shows dynamic interplays among key decision factors. Demand for the current product is determined by two forces, i.e., the product's price and its EOL warranty period. That is, the lower the price and/or the longer the warranty period, the larger the demand. The larger the demand, the more profit the firm will achieve

[4]For more theoretical explanation, refer to Tomiyama, K. (1984). 'Two-stage optimal control problems and optimality conditions.' *Journal of Economic Dynamics and Control*, **9**, 3, 317-337. & Rossana, R. J. (1985). 'Delivery lags and buffer stocks in the theory of investment by the firm.' *Journal of Economic Dynamics and Control*, **9**, 2, 153–193.

in the first stage. The firm will also gain a lump-sum profit at the end of the first stage, depending on the size of the installed base of the current product. For example, the lump-sum profit increases as the cumulative number of customers increases. There are parts failures in the first stage, which must be serviced with the spare parts manufactured in the first stage. But, the firm might want to manufacture the first-stage spare parts more than needed in the first stage in order to service the customers in the second stage. There are two conflicting forces, inventory holding costs and spare parts manufacturing cost differences: (1) Spare parts manufactured in the first stage can be used in the second stage, incurring inventory holding cost, and (2) unit cost to produce a spare part in the first stage is less than that in the second stage, implying it is economical to make a spare part while the current product is being still produced.

On the one hand, the firm wants to offer a long warranty period in order to stimulate the demand of the current product. On the other hand, the future cost to service the customers will increase as the warranty period extends. Similarly, in order to increase the current product sales, the firm might have to reduce the price, but as the cumulative sales increase, the future liability to service the customers in the second stage also increases. Therefore, the decision problem faced by the firm is not straightforward: The firm has to balance multiple forces simultaneously.

Figure 5 shows our conjecture of the dynamics of the state variables. A key presumption is t^* exists between T and $T + \tau$: We will more elaborate on this soon. Table 3 summarizes notations used in the optimal control theory model. In addition, the model is based on assumptions, which define the context of our problem:

- The product life cycle of the current product is T. That is, the company will produce and sell the current product until T.

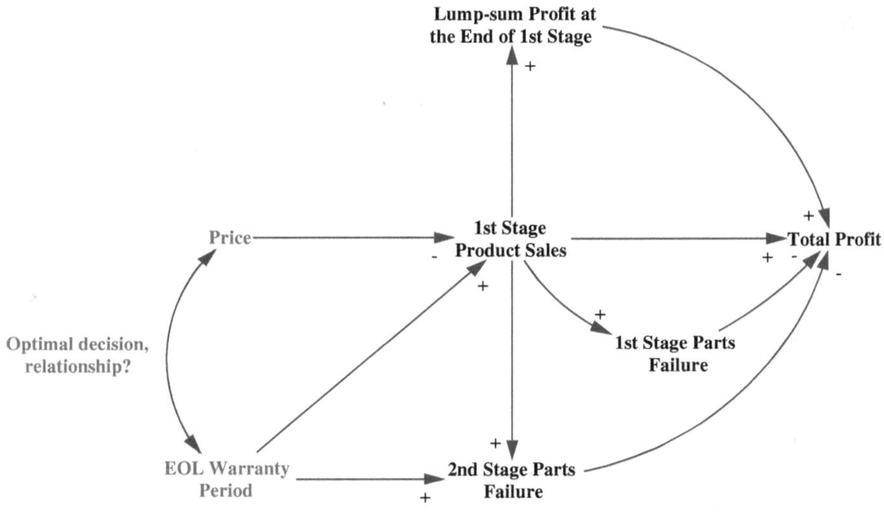

Fig. 4 Dynamics of the decision problem

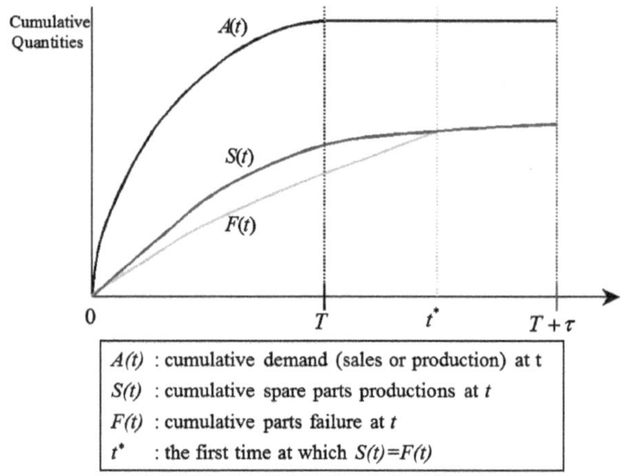

Fig. 5 Example dynamics of key state variables

Table 3 Notation

$p(t)$:	Sales price of the current product at t
$q(t)$:	Production amount of spare parts at t
c:	Unit production cost of the current product
c_i:	Unit cost to replace the spare part at the stage i, $i = 1, 2$
k_i:	Cost parameter associated with manufacturing the spare part at i, $i = 1, 2$
h:	Unit inventory cost of the spare part
$A(t)$:	Cumulative sales units of the current product at t
$S(t)$:	Cumulative production units of the spare part at t
$F(t)$:	Cumulative parts failures at t
T:	Life cycle of the current product
τ:	Warranty period
π_R:	Unit salvage value of the leftover spare part at $t = T + \tau$, i.e., $\eta(T + t) \equiv \pi_R$
$\dot{A}(t)$:	Sales, i.e., demand, of the product at t, i.e., $\dot{A}(t) = d_1 - d_2 p(t) + d_3 \tau$
d_1:	Potential market size when the price is zero and the warranty is zero
d_2:	Price coefficient in the demand function; if the price is $p(t)$, the potential demand is reduced by $d_2 p(t)$
d_3:	Warranty coefficient in the demand function; if the warranty period is τ, the potential demand is increased by $d_3 \tau$
Z:	Objective function to represent the net profit for the company throughout the entire decision time horizon, i.e., $t \in [0, T + \tau]$

- The warranty period for the key part of the product is τ. That is, the company has to provide the customer with the key spare part of the product until $T + \tau$.
- The customer can get the spare part from the company only once throughout the entire product life cycle including the warranty period. In fact, for $t \leq T$, the company guarantees to service the failed part (as long as each customer gets the

service only once), regardless of τ. That is, as long as the current product is being made, every unit is entitled to get serviced once, if necessary.

- The company must honor its warranty, i.e., it must be that $S(t) \geq F(t)$ throughout the decision time horizon.
- The rate of parts failure is constant, e.g., the number of parts failed at t is $\alpha(A(t) - F(t))$, where $0 < \alpha < 1$ is a constant. This assumption is compatible with the memory-less property of a renewal process. That is, the instantaneous probability that a working product breaks down at t is always the same, i.e. α.
- The production lead time is sufficiently short, e.g., there will be no inventory of the product. There is only the inventory of spare parts.
- The warranty period affects the demand in an additive way, i.e., $\dot{A}(t) = d_1 - d_2 p(t) + d_3 u(\tau)$, where $u(\tau) = \tau$. Therefore, as the warranty period extends, the potential market size enlarges, i.e., from d_1 to $d_1 + d_3\tau$. There are other forms of demand function, e.g., a multiplicative rather than an additive relationship: An example of such a multiplicative relationship is $\dot{A}(t) = (d_1 - d_2 p)(\tau + d_4)^{d_3}$, where the warranty period, τ, affects the demand in a nonlinear way. After trying to adopt different functional forms, we have concluded that the analysis outcomes are qualitatively similar. Thus, we focus only on the additive form in the ensuing analysis.

Employing the notations, we set up a two-stage optimal control theory model:

$$Max\,Z = \int_0^T \left[(p(t) - c)(d_1 - d_2 p(t) + d_3\tau) - h\{S(t) - F(t)\} - \frac{k_1}{2} q(t)^2 - c_1\{\alpha(A(t) - F(t))\} \right] \mathrm{d}t + \int_T^{T+\tau} \left[-h\{S(t) - F(t)\} - \frac{k_2}{2} q(t)^2 - c_2\{\alpha(A(T) - F(t))\} \right] \mathrm{d}t + mA(T) - nA(T)^2 \quad (40)$$

Subject to

Constraints for Stage 1	Constraints for Stage 2
$\dot{A}(t) = d_1 - d_2 p(t) + d_3\tau$, (41) $\dot{S}(t) = q(t)$, (42) $\dot{F}(t) = \alpha\{A(t) - F(t)\}$, (43) $S(t) \geq F(t)$, (44) $S(0) = 0,\ A(0) = 0,\ F(0) = 0.$ (45) $A(T)\,free$ (46)	$\dot{A}(t) = 0$ (47) $\dot{S}(t) = q(t)$, (48) $\dot{F}(t) = \alpha\{A(T) - F(t)\}$, (49) $S(t) \geq F(t)$ (50)

The objective function consists of 3 major parts, i.e., the profit function at the first stage for $0 \leq t \leq T$, the cost function at the second stage for $T \leq t \leq T + \tau$, and the lump sum profit based on the first stage's cumulative sales (installed base) at the end of the first stage $t = T$.

The integrand of the first stage's profit function is comprised of 4 elements:

- $(p(t)-c)(d_1-d_2p(t)+d_3\tau)$: Since $(p(t)-c)$ is the unit contribution and $(d_1-d_2p(t)+d_3\tau)$ sales volume at t, $(p(t)-c)(d_1-d_2p(t)+d_3\tau)$ is the total profit (contribution) from the sales of the current product at t.
- $h\{S(t)-F(t)\}$: Since $\{S(t)-F(t)\}$ is the amount of spare parts that are not consumed by t and h, the inventory management cost for a unit spare part per period, $h\{S(t)-F(t)\}$, is the total inventory management cost of the spare parts at t.
- $\frac{k_1}{2}q(t)^2$: It is the total production cost to manufacture $q(t)$ units of spare part at t.
- $c_1\{\alpha(A(t)-F(t))\}$: Since $\alpha(A(t)-F(t))$ is the total number of part failures at t and c_1, the unit cost to service the failed part, $c_1\{\alpha(A(t)-F(t))\}$, is the total cost to service the spare parts failures at t.

The integrand of the cost function at the second stage consists of elements similar with those for the first stage, except for that in the second stage, there is no contribution from the sales of the current product. In addition, we assume that $k_1 < k_2$ and $c_1 < c_2$, implying that it is more economical to make a spare part during the first stage than during the second, and also, it is less expensive to service a spare part during the first stage than during the second. It is reasonable because during the second stage, the company no longer makes the product, and therefore, the spare parts manufacturing enjoys neither economies of scale nor economies of scope.

$mA(T)-nA(T)^2 \equiv RS$: $A(T)$ is the total cumulative number of current products sold during the first stage. We assume that the company can benefit from the installed base of the current product at the end of the first stage. For example, the larger the installed base at T, the higher the firm's brand value, which in turn helps the company to sell new products more effectively or in larger quantity. $mA(T)-nA(T)^2$ represents such a lump sum profit the company garners from the total sales during the first stage. Normally, it would be expected that the lump sum profit increases in $A(T)$, i.e., $\frac{d(RS)}{dA(T)} > 0$, but the rate of increasing is decreasing, i.e., $\frac{d^2(RS)}{dA(T)^2} < 0$.

We have different sets of constraints for the two stages, (41)–(46) for the first stage and (47)–(50) for the second stage. (41) is the demand for the current product at t, (42) represents the amount of spare parts manufactured at t, (43) is the amount of spare parts failures, (44) is the constraint to make sure that the company always should be able to satisfy the spare parts demand, (45) are the initial conditions for the state variables, and (46) is the terminal condition for the cumulative sales. Now, consider the constraints for the second stage. (48) and (50) are exactly the same as (42) and (44). (47) implies that the company no longer manufactures and sells the product during the second stage, and (49) is the same as (43) except for $A(T)$ instead of $A(t)$ since there will be no change in the cumulative sales volume in the second stage.

Recall that t^* is the first time when $S(t) = F(t)$ occurs, which implies that the spare parts produced are consumed completely by t^*, and after t^*, the company has to make the spare parts in the same amount as needed. In Fig. 5, we divided the entire life cycle (i.e.,$T + \tau$) into three periods, assuming t^* exists such as $T < t^* < T + \tau$:

(1) Stage 1: $0 \le t \le T$, during which the company manufactures and sells the current product.
(2) Stage 2–1: $T \le t \le t^*$, from the time the production of the current product is discontinued to the time at which the cumulative parts production equals the cumulative parts failure.
(3) Stage 2–2: $t^* \le t \le T + \tau$: from t^* to the end of the entire life cycle including the warranty period.

$T < t^* < T + \tau$ is not a very stringent condition. Since we assume $k_1 < k_2$, it is better for the company to make spare parts more than needed during the first stage, and therefore, it is likely that $T < t^*$; the unit inventory holding cost should not be too large. In addition, we assume that the salvage value of a leftover spare part at $T + \tau$ is lower than the unit spare part manufacturing cost during the second stage, i.e., $c_2 > \pi_R$: This assumption is necessary and realistic, since otherwise the company might be tempted to produce unlimited number of spare parts, any left-over of which can be sold with profit at $T + \tau$. As a result, it is reasonable to assume $t^* < T + \tau$; thus, $T < t^* < T + \tau$. Even if we do not impose $T < t^* < T + \tau$, the problem-solving procedure is similar, the difference being that it involves more complicated steps without generating better insights. In the ensuing discussion, we adopt the condition $T < t^* < T + \tau$ as given.

Theorem 4 *Solutions that satisfy the necessary conditions are derived as follows. For Stage 1, i.e.,* $0 \le t \le T$,

$$p^*(t) = \frac{1}{2}\left(\frac{d_1}{d_2} + c + \frac{d_3\tau}{d_2} - X_3\right) + \frac{h}{2}t + \frac{X_2}{2}e^{\alpha t}, \tag{51}$$

$$q^*(t) = \frac{h}{k_1}t + \frac{X_1}{k_1}, \tag{52}$$

$$A^*(t) = \frac{1}{2}\left\{(d_1 - cd_2 + d_2X_3 + d_3\tau)t - \frac{d_2h}{2}t^2 - \frac{d_2X_2}{\alpha}e^{\alpha t}\right\} + \frac{d_2X_2}{2\alpha}, \tag{53}$$

$$F^*(t) = \frac{1}{2}\left(\frac{cd_2 + d_2X_2 - d_1 - d_2X_3 - d_3\tau}{\alpha} - \frac{d_2h}{\alpha^2}\right) + \frac{t}{2}\left(d_1 - cd_2 + d_2X_3 + d_3\tau + \frac{d_2h}{\alpha}\right) - \frac{d_2X_2e^{\alpha t}}{4\alpha} - \frac{d_2ht^2}{4} + \left[\frac{d_2X_2}{4\alpha} - \frac{1}{2}\left(\frac{cd_2 + d_2X_2 - d_1 - d_2X_3 - d_3\tau}{\alpha} - \frac{d_2h}{\alpha^2}\right)\right]e^{-\alpha t}, \tag{54}$$

$$S^*(t) = \frac{h}{2k_1}t^2 + \frac{X_1}{k_1}t. \tag{55}$$

For Stage 2–1, i.e., $\boldsymbol{T \leq t \leq t^*}$,

$$A(t) = \frac{1}{2}\left[(d_1 - d_2c + d_2X_3 + d_3\tau)T - \frac{d_2h}{2}T^2 - \frac{d_2X_2}{\alpha}e^{\alpha T}\right] + \frac{d_2X_2}{2\alpha}, \tag{56}$$

$$\begin{aligned} F^*(t) &= \frac{1}{2}\left[(d_1 - d_2c + d_2X_3 + d_3\tau)T - \frac{d_2h}{2}T^2 - \frac{d_2X_2}{\alpha}e^{\alpha T}\right] + \frac{d_2X_2}{2\alpha} \\ &+ \left[\frac{1}{2}\left(\frac{cd_2 - d_1 - d_2X_3 - d_3\tau}{\alpha} - \frac{d_2h}{\alpha^2}\right)e^{\alpha T} + \frac{d_2h}{2\alpha}e^{\alpha T}T + \frac{d_2X_2}{4\alpha}e^{2\alpha T} + \left[\frac{d_2X_2}{4\alpha} - \frac{1}{2}\left(\frac{cd_2 - d_1 - d_2X_3 - d_3\tau}{\alpha} - \frac{d_2h}{\alpha^2}\right)\right]\right]e^{-\alpha t}, \end{aligned} \tag{57}$$

$$S^*(t) = \frac{h}{2k_2}t^2 + \frac{X_1}{k_2}t + \left(\frac{1}{k_1} - \frac{1}{k_2}\right)\left(\frac{hT^2}{2} + X_1T\right). \tag{58}$$

For Stage 2–2, i.e., $\boldsymbol{t^* \leq t \leq T + \tau}$,

$$\begin{aligned} F^*(t) &= S^*(t) = A(T) + \frac{h - \pi_R}{\alpha^2 k_2}e^{\alpha(T + \tau - \alpha t)} \\ &= \frac{1}{2}\left[(d_1 - d_2c + d_2X_3 + d_3\tau)T - \frac{d_2h}{2}T^2 - \frac{d_2X_2}{\alpha}e^{\alpha T}\right] + \frac{d_2X_2}{2\alpha} + \frac{h - \pi_R}{\alpha^2 k_2}e^{\alpha(T + \tau - \alpha t)}. \end{aligned} \tag{59}$$

The three integration constants are determined as follows.

$$X_1 = -ht^* - \frac{(h - \pi_R)}{\alpha}e^{\alpha(T + \tau - t^*)}. \tag{60}$$

The relationship between X_2 and X_3 is defined in the following two equations, (61) and (62).

$$X_3 = \frac{8}{1}(1 - nd_2T)\left[n\left\{(d_1 - d_2c + d_3\tau)T - \frac{d_2h}{2}T^2 - \frac{d_2X_2}{\alpha}e^{\alpha T}\right\} + \frac{nd_2X_2}{\alpha} + X_2e^{\alpha T} - m + hT\right], \tag{61}$$

$$\begin{aligned} &\frac{1}{2}\left(\frac{cd_2 - d_1 + d_2X_2 - d_2X_3 - d_3\tau}{\alpha} - \frac{d_2h}{\alpha^2}\right)e^{\alpha T} + \frac{d_2h}{2\alpha}e^{\alpha T}T + \frac{d_2X_2}{4\alpha}e^{2\alpha T} - \frac{d_2X_2}{2\alpha}e^{\alpha T} \\ &+ \left[\frac{d_2X_2}{4\alpha} - \frac{1}{2}\left(\frac{cd_2 - d_1 + d_2X_2 - d_2X_3 - d_3\tau}{\alpha} - \frac{d_2h}{\alpha^2}\right)\right] = \frac{h - \pi_R}{\alpha^2 k_2}e^{\alpha(T + \tau)}. \end{aligned} \tag{62}$$

Proof The necessary conditions are as follows.

For Stage 1, the Hamiltonian is (63) and its necessary conditions are (64)–(70).

$$H_1 = (p(t) - c)(d_1 - d_2p(t) + d_3\tau) - h\{S(t) - F(t)\} - \frac{k_1}{2}q(t)^2 - c_1\{\alpha(A(t) - F(t))\} + \lambda(d_1 - d_2p(t) + d_3\tau) + \gamma q(t) + \theta\alpha\{A(t) - F(t)\} + \eta\{S(t) - F(t)\}, \tag{63}$$

$$\frac{\partial H_1}{\partial p} = (d_1 + cd_2 + d_3\tau) - 2d_2p - \lambda d_2 = 0, \tag{64}$$

$$\frac{\partial H_1}{\partial q} = -k_1q(t) + \gamma = 0, \tag{65}$$

$$\lambda' = -\frac{\partial H_1}{\partial A} = c_1\alpha - \alpha\theta, \tag{66}$$

$$\gamma' = -\frac{\partial H_1}{\partial S} = h - \eta, \tag{67}$$

$$\theta' = -\frac{\partial H_1}{\partial F} = \alpha\theta + \eta - c_1\alpha - h, \tag{68}$$

$$\eta\{S(t) - F(t)\} = 0,\ \eta \geq 0,\ S(t) - F(t) \geq 0, \tag{69}$$

$$\lambda(T) = 0. \tag{70}$$

For Stage 2, the Hamiltonian is (71) and its necessary conditions are (72)–(75).

$$H_2 = -h\{S(t) - F(t)\} - \frac{k_2}{2}q(t)^2 - c_2\{\alpha(A(T) - F(t))\} + \gamma q(t) + \theta\alpha\{A(T) - F(t)\} + \eta\{S(t) - F(t)\}, \tag{71}$$

$$\frac{\partial H_2}{\partial q} = -k_2q(t) + \gamma = 0, \tag{72}$$

$$\gamma' = -\frac{\partial H_2}{\partial S} = h - \eta, \tag{73}$$

$$\theta' = -\frac{\partial H_2}{\partial F} = \alpha\theta + \eta - c_2\alpha - h, \tag{74}$$

$$\eta\{S(t) - F(t)\} = 0,\ \eta \geq 0,\ S(t) - F(t) \geq 0. \tag{75}$$

Other additional conditions as switching conditions at $t = T$ are (76)–(79).

$$H_1(T) + R_T(T) = H_2(T), \tag{76}$$

$$\lim_{t\to T^-} \gamma(t) = \lim_{t\to T^+} \gamma(t), \tag{77}$$

$$\lim_{t\to T^-} \theta(t) = \lim_{t\to T^+} \theta(t), \tag{78}$$

$$\lambda^-(T) + m - 2nA(T) = \lambda^+(T). \tag{79}$$

Using the necessary conditions as in (63)–(79), we can derive the solutions in (51)–(62) that satisfy the necessary conditions. For details, see Appendix 2.

Theorem 5 *The solutions that satisfy the necessary conditions in Theorem 4 are optimal.*

Proof We first show that the objective function is concave jointly in p and q and that the constraints are linear in p and q.

Partially, twice-differentiating the integrand in the first part of (1) with regard to p and q, we get:

$$\begin{aligned}\frac{\partial^2}{\partial^2 p}&\{(p(t)-c)(d_1-d_2p(t)+d_3\tau)-h\{S(t)-F(t)\}\\&-\frac{k_1}{2}q(t)^2-c_1\{\alpha(A(t)-F(t))\}\}=-2d_2<0,\end{aligned} \tag{80}$$

$$\begin{aligned}\frac{\partial^2}{\partial^2 q}&\{(p(t)-c)(d_1-d_2p(t)+d_3\tau)-h\{S(t)-F(t)\}\\&-\frac{k_1}{2}q(t)^2-c_1\{\alpha(A(t)-F(t))\}\}=-k_1<0.\end{aligned} \tag{81}$$

Likewise, we partially twice-differentiate the integrand in the second part of (1) with regard to p and q, and we get:

$$\frac{\partial^2}{\partial^2 p}\{-h\{S(t)-F(t)\}-\frac{k_2}{2}q(t)^2-c_2\{\alpha(A(T)-F(t))\}\}=0, \tag{82}$$

$$\frac{\partial^2}{\partial^2 q}\{-h\{S(t)-F(t)\}-\frac{k_2}{2}q(t)^2-c_2\{\alpha(A(T)-F(t))\}\}=-k_2<0. \tag{83}$$

Therefore, (80)–(83) prove that the objective function is concave in p and q. In addition, we partially twice-differentiate the constraints (41)–(50) and all of the values are zero, implying that the constraints are all linear in p and q.

In conclusion, the necessary conditions in Theorem 4 are also sufficient conditions for optimality.

Theorem 6 *Given T, the optimal price is always decreasing.*

Proof From (64), we know $p^*(t) = \frac{d_1 + cd_2 + d_3\tau - \lambda d_2}{2d_2}$.

Therefore,

$$\dot{p} = \frac{d}{dt}\left\{\frac{d_1 + cd_2 + d_3\tau - \lambda d_2}{2d_2}\right\} = -\frac{\dot{\lambda}}{2}. \tag{84}$$

From (66), $\dot{\lambda} = c_1\alpha - \alpha\theta$. θ is costate variable associated with $F(t)$, which is the cumulative number of spare part's failures. From an economic sense, θ represents the shadow price of $F(t)$. Thus, it is reasonable to conclude that $\theta < 0$ always, i.e., $0 \le t \le T$, since the larger the cumulative number of part's failures, the less the total profit and $\dot{F} \ge 0$. Now that $\theta < 0$ and $\dot{\lambda} > 0$ from (66) and $\dot{p} < 0$ for $0 \le t \le T$ from (84).

Theorem 6 implies that the company controls its demand throughout the product life cycle: During the early product life cycle, the company tries to suppress the demand by charging higher prices, but as the product life cycle approaches the end, the company increases the demand by lowering the prices. What would be the motivations behind such a pricing strategy? There are multiple forces that have interacted with each other to determine the optimal dynamics. First, there is an inventory holding cost associated with the spare part. In order to avoid the inventory management cost, the firm has an incentive to suppress the demand during the early product life cycle so as to minimize the need to make the spare parts early. Second, there is a lump sum payoff that is based on the cumulative sales volume at the end of the product life cycle. In order to maximize the lump sum payoff, the company has an incentive to increase the demand. By taking into account the need to reduce the inventory holding cost, the company tries to increase the demand as the product life cycle approaches the end and thus reduces the price sharply.

3.2 Numerical Analysis

In order to see and understand the patterns of key variables' optimal dynamics, we conduct numerical analysis. Parameter values used in the analysis are in Table 4. These values are based on a real-world case in the mobile phone manufacturing industry, briefly mentioned in the introduction section. The case company offers the following information: (1) It adopts a price-skimming strategy, starting from about

Table 4 Parameter values for the numerical analysis

c	c_1	c_2	d_1	d_2	d_3	h	k_1	k_2	α	π_R	m	n
4	1	2	30	1	0.1	0.5	1	2	0.1	1.0	100	0.5

\$300 initially and decreasing to the level almost equivalent to the production cost, (2) the unit production cost of a mobile phone is about \$40 ($c = 4$), the unit cost to service a failed unit is about \$10, while the phone is still in production ($c_1 = 1$) or \$20 if the company no longer makes the phone in-house ($c_2 = 2$), (3) other cost parameters such as unit production cost, unit inventory holding cost, and unit salvage value of spare parts are consistent in magnitude with the production and servicing costs, (4) the potential market size is about 300,000 units per month; thus, $d_1 = 30$, and the price and warranty coefficients are estimated by the company, (5) the failure rate is about 10%, i.e., at a given time, 10% of the units currently working in the market fail, requiring the replacement of the key part, and (6) other parameter values are all consistent with the company's operations data.

There are important dynamic relationships derived from the numerical analysis. We first look into the relationship between optimal T and τ. Since we allow T to vary, we need to determine an optimal T first and then to calculate τ given the optimal T. We also present the optimal dynamics of state and control variables, which are consistent with those hypothesized in Fig. 5. Finally, we show several sensitivity analyses associated with optimal dynamic interaction among price, part failure rate, warranty period, and product life cycle.

3.2.1 The Optimal Warranty Period for a Given *T*

As the product life cycle increases, the optimal warranty period decreases up to a point and after that point increases again: This pattern is depicted in Fig. 6. What forces cause this phenomenon to occur? When the product life cycle is relatively short, the firm needs to offer relatively extensive EOL warranty in order to attract more customers so as to generate enough sales during the short product life. As the product life lengthens, however, the incentive to offer extensive EOL warranty to attract customers lessens (because the firm now can generate enough sales 'slowly' during the 'extended' product life cycle), and thus, the firm can afford to reduce the warranty period. But, this is true only up to a point. That is, once the product life extends past the critical point, the firm has to increase the EOL warranty period in order to attract demand that can last long enough for the 'extensive' product life cycle (an extended warranty will enlarge the market potential). The important question is how to identify the critical point. In our numerical example, the total profit is maximized at $T = 11.6$, at which the warranty period is minimized as well. Is it always true that the total profit is maximized when the warranty period is minimized? It is an intriguing, but difficult question to answer given our present analysis. As such, it can be an important future research question.

3.2.2 Optimal Dynamics of State and Control Variables

The dynamics of control and state variables are consistent with what we suggested in Fig. 5. Figure 7 shows the optimal patterns of cumulative total sales unit,

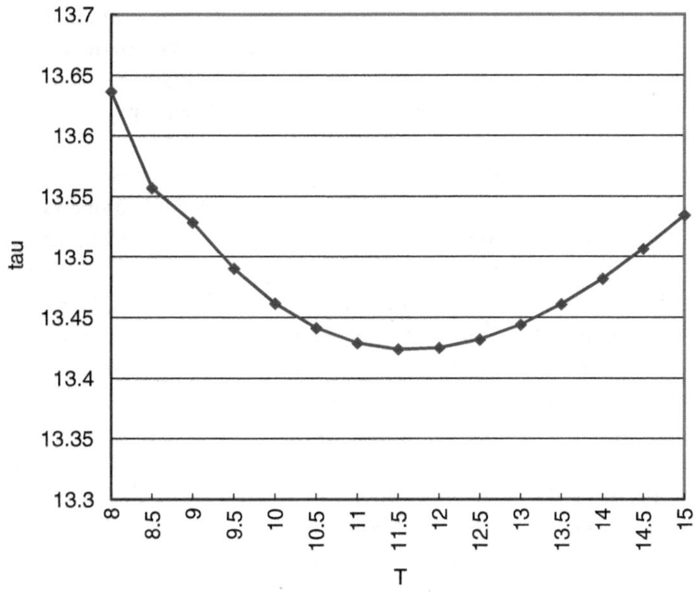

Fig. 6 Optimal relationship between warranty period (τ) and T

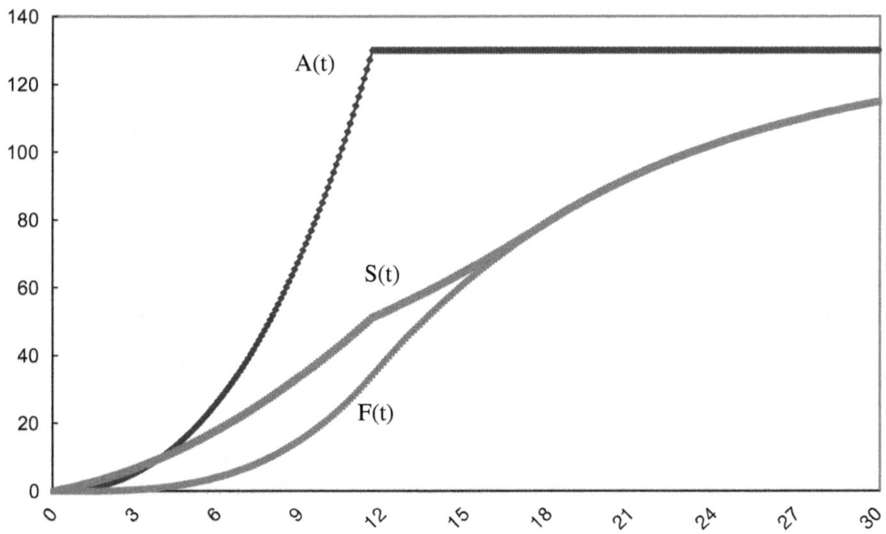

Fig. 7 Optimal dynamics of state variables

cumulative parts production, and cumulative parts failure. Given the parameter values in Table 4, the firm's optimum solution would be to make and sell the product until $T = 11.6$ with warranty for about 15.4 time period, i.e., $\tau = 15.4$. Also, at $t^* = 17.73$, the optimal cumulative number of parts production equals the cumulative parts failure.

Theorem 6 about the optimal dynamics of pricing is confirmed by Fig. 8. That is, the optimal price is always decreasing. Figure 9 shows the dynamic patterns of spare parts production. The optimal decision for the firm would be to increase the spare parts production gradually until the end of product life cycle, T, at which the optimal production amount precipitates in a sudden disruption. Then, the production increases again, albeit the rate is slower than in the first stage, until t^* is reached. After t^*, the production amount coincides with the spare parts failures. These dynamics can be explained by the difference in spare parts production costs between the two stages. That is, since we assume it is more expensive to make spare parts in the second stage than in the first stage, the company has an incentive to make spare parts in the first stage more than needed in that stage. But, there involves another force, the inventory holding cost. As a result, although the company tries to make spare parts more than needed in the first stage, it also wants to delay making spare parts in order to avoid inventory cost as much as possible. In Fig. 9, we see the company increases its production rate for spare parts very steeply in the first stage. The remaining pattern of spare parts manufacturing in the second stage can be characterized as firm's effort to manage the production quantity so as to minimize the leftover spare parts in the end.

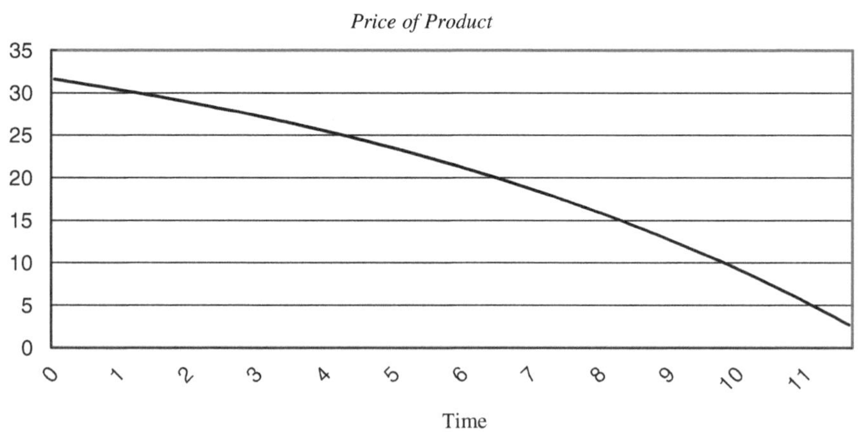

Fig. 8 Optimal pricing dynamics

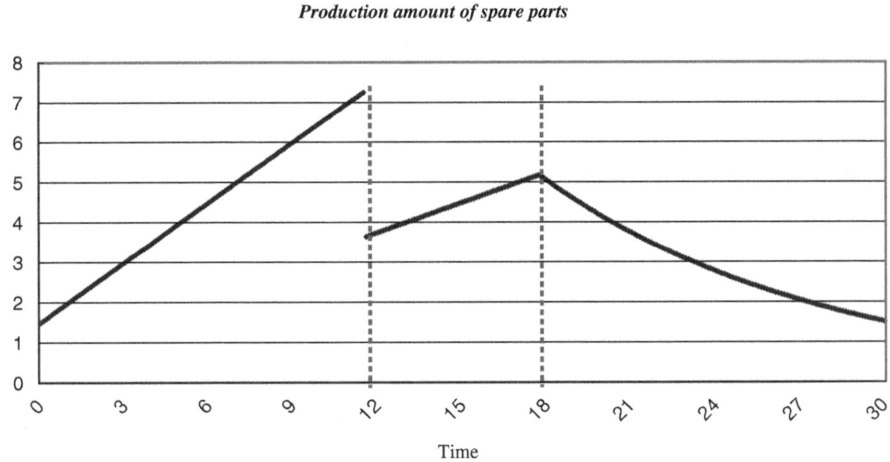

Fig. 9 Optimal dynamics of spare parts manufacturing

3.2.3 Sensitivity Analysis and Failure Rate Changes

In this section, we focus on how the failure rate (α) changes affect optimal warranty period, total profit, and optimal pricing dynamics. In order to observe the dynamics more clearly, we need to use a fixed product life cycle: For the ensuing analysis, we use the base model's product life cycle, i.e., $T^* = 11.6$.

Figure 10 shows that as the failure rate increases, the optimal warranty period increases up to a point and then decreases. More specifically, in our example, the optimal warranty period peaks at $\alpha = 0.120$, after which it decreases: The longest warranty period is about $\tau = 13.74$. In essence, the failure rate α represents the defect rate: The larger the failure rate, the lower the quality level, i.e., the higher the defect rate. Therefore, the dynamics in Fig. 10 imply that as the defect rate increases, the company has to increase the warranty period so as to compensate for the deteriorating quality level, but only up to a certain point. We can imagine a threshold level of defect rate, after which the company should start decreasing the warranty since the cost to service all the failed parts becomes prohibitive when the failure rate is excessively high, and thus, the company wants to reduce the service costs by curtailing its warranty to the customers. This explanation is consistent with the relationship between total profit and failure rate in Fig. 11, where the company's total profit monotonically decreases as the failure rate increases.

Figure 12 enables us to infer how the optimal pricing dynamics change as the defect rate changes. When the failure rate is low, the company enjoys more latitude to determine optimal prices, implying that it can charge higher prices in the early period of the product life cycle.

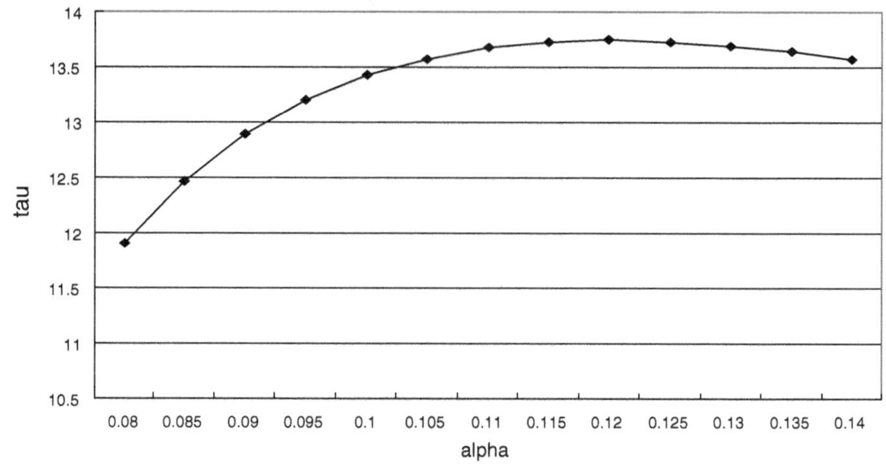

Fig. 10 Failure rate and warranty period

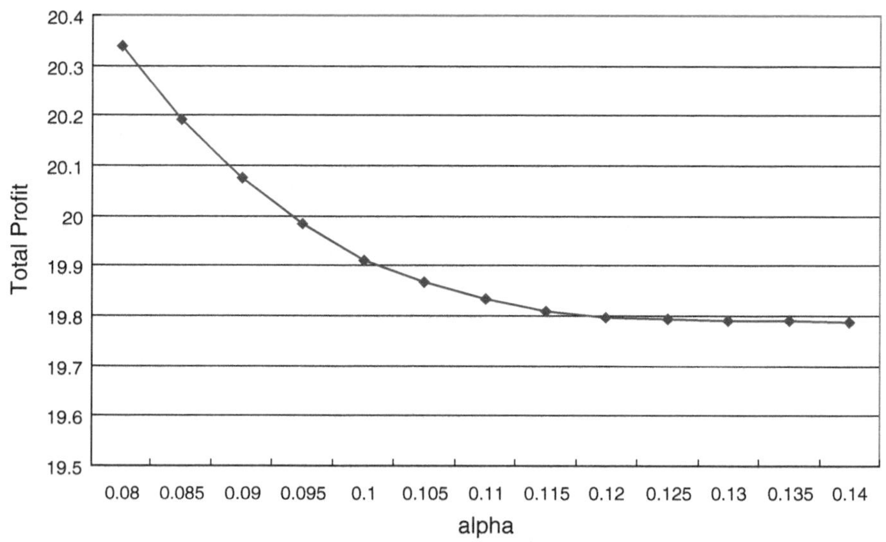

Fig. 11 Failure rate and total profit

3.3 Managerial Implications

We started with a primary question focused on the dynamic relationship among firm's EOL warranty period, optimal pricing, and spare parts production. In order to answer the question, we developed a two-stage optimal control theory model, where the first stage is the period during which the firm makes and sells the product and

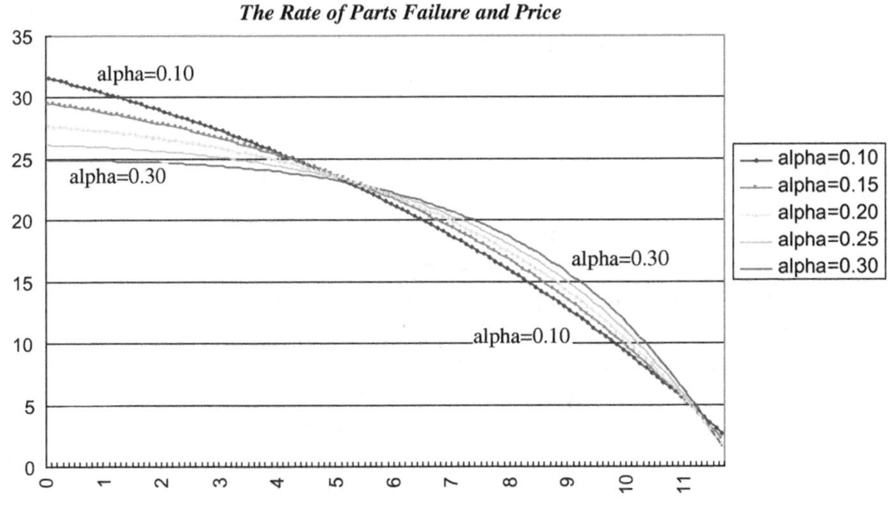

Fig. 12 Failure rate and pricing

the second stage consists of the EOL warranty period during which the firm has to honor its commitment to servicing the product with the spare part.

The two-stage optimal control theory model was complex so as to make it practically unappealing to derive closed-form solutions. Therefore, rather than employing general functional forms, we incorporated into the model specific ones, each being based on more realistic assumptions that were extensively utilized in the literature. Key assumptions of the model are as follows:

- EOL warranty affects the demand function in an additive manner. That is, the longer the EOL warranty period, the larger the demand for the product (Wiener 1985). See Eq. (41).
- At the end of the current product's life cycle, the company earns lump-sum profit depending on the cumulative sales units of the product (Thompson 1968). That is, the larger the installed base at the end of the first stage, the larger the lump-sum benefit. Amount of the lump-sum profit is concave in the size of the cumulative sales units.
- The cost function to make spare parts is quadratic in the production amount (Eliashberg and Steinberg 1991). Other cost functions such as inventory holding cost and spare part servicing cost are all linear. This assumption is consistent with the actual situation faced by the case company in the numerical example. The company needs special capacity (in addition to the regular capacity to make the finished phone itself) to manufacture the spare part, forcing the total cost of manufacturing spare parts to behave in a quadratic, i.e., concave and increasing, pattern. That is, the marginal cost to make an additional spare part is affected by the scale effect in the capacity: The more the company manufacturers the spare part, the larger the marginal cost. Unlike manufacturing spare parts, other

normal operations activities such as making the phones, managing inventory, and servicing failed units are accommodated by the regular capacity in the company: Thus, the cost functions related with these normal operations behave in linear patterns.

- Spare parts failure follows a deterministic pattern, which in essence is comparable to a memory-less property (Barros 1989). That is, a constant portion of 'working (i.e., not-yet-broken)' products fails at a given moment.
- The company's service failure is not acceptable. That is, the company must guarantee servicing customers with spare parts at most once during the product life cycle. In order to be competitive in the market, this kind of commitment is not unusual nowadays.
- Production lead time for both the current product and its spare part is assumed to be negligible. That is, the production lead time is very short compared with the overall decision-making speed so that it practically does not affect the decision process. Similarly, the product is in great demand, i.e., as soon as it is made, it is sold rapidly in the market. In essence, the delivery lead time is also negligible, and thus, the production period is equal to the sales period. It is a common phenomenon, in particular, when the product has a short product life cycle and its production lead time is insignificant, and also, the product is an innovative one and its demand is larger than its supply in general. These observations are consistent with the case of the mobile phone manufacturing industry, on which we based our numerical analysis.

Each one of these assumptions is reasonable and seems to reflect the reality sufficiently. Analyzing the model based on these assumptions, we have obtained important findings.

- There exist optimal product life cycle and optimal EOL warranty period. In fact, in order to maximize total profit, the company has to decide *how long it makes and sells the current product* and *how long its EOL warranty period should be* simultaneously. For our specific model context, relationship between optimal product life cycle and optimal EOL warranty period is convex: When the product life cycle is relative short, the company can afford to curtail its EOL warranty as the product life cycle lengthens, but this pattern is valid only up to a point, that is, when the product life cycle is long enough, the company need increase its EOL warranty as the product life cycle lengthens. What forces drive this phenomenon? When the product life cycle is very short, the firm offers relatively extensive EOL warranty to attract more customers so as to generate enough sales during the short product life *quickly*. As the product life gets longer, however, the incentive to offer extensive EOL warranty to attract customers lessens since the firm now can generate enough sales 'slowly' during the 'extended' product life cycle; thus, the firm can afford to curtail its EOL warranty, but only up to a point. That is, once the product life extends beyond the threshold level, the firm has to increase its EOL warranty period in order to attract demand that can last long enough for the 'extensive' product life cycle.

- Relationship between optimal EOL warranty period and failure rate is concave: When the defect (failure) rate is moderate, the company has to increase its EOL warranty period as the defect rate increases so as to compensate for the deteriorating quality, but when the defect rate is beyond a threshold level, the company needs to curtail its EOL warranty commitment as the defect rate increases in order to avoid excessive cost to service the failed parts.

In addition, we examined other dynamics among key variables such as optimal price, price and EOL warranty elasticity of market demand, and failure rate. By employing both analytic solution and numerical analysis, we were able to depict crucial dynamics in our model. As such, we hope to shed light on how to make decision for optimal pricing and EOL warranty when the product life cycle is finite and the company is obliged to provide after-sales services to customers for an extended period of time after the current product is no longer manufactured.

Exercise Problems

1. Assume a supply chain in which a manufacturer produces a product and sells it to a retailer, who then sells it to the final market. A retailer coordinates, e.g., supports, a manufacturer's quality innovation that can lead to the improvement of its product quality. The retailer invests in u units of the manufacturer's quality innovation projects, which would cost $\rho u^2/2$ to the retailer. The product quality increases with the innovation projects according to $\dot{Q} = u$. The market demand is dependent on the price and the product quality, i.e., $D = \alpha - \beta p + \gamma Q$. The sales price p and the wholesale price w are exogenous. Manufacturer's unit production cost is c. In addition to the production cost, a deviation from the manufacturer's effective capacity $\overline{U}$ incurs a penalty cost to the manufacturer, i.e., $c_d\left(D - \overline{U}\right)^2$. A retailer decides $u(t)$ to maximize its profit over the fixed planning period $0 \leq t \leq T$. We assume a short enough T so that the discounting effect can be disregarded and the initial quality level at $t = 0$ is Q_0, i.e., $Q(0) = Q_0$.
   (a) Develop an optimal control problem for the retailer.
   (b) Determine the optimal coordination strategy $u(t)$ for the retailer.
2. Suppose that a supplier and a manufacturer invest u_i in innovation capability x_i. Innovation capability of a firm leads to quality innovation of the product, thereby increases the consumer demand from D to $D + k(x_1 + x_2)$. The parameters are described below.

D	Initial demand (constant)
p_1	Wholesale price (constant)
p_2	Sales price (constant)
c_1	Supplier's internal unit production cost (constant)
c_2	Manufacturer's internal unit production cost (constant)
k	Coefficient associated with the demand increase by innovation
x_1	Supplier's cumulative innovation capability
x_2	Manufacturer's cumulative innovation capability
α_1	Coefficient associated with supplier's investment cost (constant)
α_2	Coefficient associated with manufacturer's investment cost (constant)
u_1	Supplier's investment in innovation capability at t
u_2	Manufacturer's investment in innovation capability at t
r	Discount rate, $0<r<1$

Suggest the open-loop Nash equilibrium strategies of the differential game problem given as follows:

$$\begin{aligned} Max_{u_1}\, J^s &= \int_0^{\infty} e^{-rt}[(D+k(x_1+x_2))(p_1-c_1)-\alpha_1 u_1-\frac{1}{2}u_1^2]\mathrm{d}t \\ Max_{u_2}\, J^m &= \int_0^{\infty} e^{-rt}[(D+k(x_1+x_2))(p_2-p_1-c_2)-\alpha_2 u_2-\frac{1}{2}u_2^2]\mathrm{d}t \\ Subject\,to\, \dot{x}_1 &= u_1+x_1,\ x_1(0)=x_{10},\ \dot{x}_2=u_2+x_2, x_2(0)=x_{20} \end{aligned}$$

3. In a supply chain consisting of a manufacturer and a retailer, suppose that there are two ways to increase the demand of a product: (i) by increasing the retailer's short-term advertising effort $A(t)$ and (ii) by increasing the long-term reputation of the product $G(t)$. Specifically, demand is given as $D(t)=D_0+\rho A(t)+\theta G(t)$, where D_0 is a base demand. Product reputation evolves according to $\dot{G}(t)=u(t)-\delta G(t)$ and $G(0)=G_0$, that is, it increases with the innovation effort of the manufacturer $u(t)$, but decays over time. Assume that each firm's unit margins are fixed and let π_m and π_r denote the unit margins of the manufacturer and the retailer, respectively. Also, assume that the advertising effort incurs a quadratic cost $A(t)^2$ to the retailer and the innovation effort incurs a quadratic cost $u(t)^2$ to the manufacturer. The retailer wants to determine the optimal advertising effort, and the manufacturer wants to determine the optimal innovation effort over $t \in [0, T]$ to maximize their respective profit.
   (a) Develop a differential game problem.
   (b) Suggest the feedback Nash equilibrium strategies for the retailer and the manufacturer.

Appendix 1

Derivation of Theorem 1

Consider (1)–(3). We impose $a_i + k_i(p_{3-i} - p_i) \geq 0$ and assume $(\rho_1 L_1 + \rho_2 L_2) - (x_1 + x_2) \geq 0$.

With the Hamiltonian

$$H_i = (p_i - c_i)((\rho_1 L_1 + \rho_2 L_2) - (x_1 + x_2))(a_i + k_i(p_{3-i} - p_i)) - \phi_i L_i^2 + \lambda_{i,i}((\rho_1 L_1 + \rho_2 L_2) - (x_1 + x_2))(a_i + k_i(p_{3-i} - p_i)) + \lambda_{i,3-i}((\rho_1 L_1 + \rho_2 L_2) - (x_1 + x_2))(a_{3-i} + k_{3-i}(p_i - p_{3-i})) \quad (85)$$

We get,

$$H^1_{p_1} = (a_1 + k_1(p_2 - p_1)) - k_1(p_1 - c_1) - k_1\lambda_{11} + k_2\lambda_{12} = 0 \quad (86)$$

and

$$H^2_{p_2} = (a_2 + k_2(p_1 - p_2)) - k_2(p_2 - c_2) + k_1\lambda_{21} - k_2\lambda_{22} = 0 \quad (87)$$

Using (86) and (87), we obtain

$$p_i = \frac{1}{3k_i k_{3-i}}(2k_{3-i}a_i - k_i k_{3-i}\lambda_{3-i,3-i} + k_i k_{3-i}c_{3-i} - 2k_i k_{3-i}\lambda_{ii} + k_i a_{3-i} + 2k_{3-i}^2\lambda_{i,3-i} + 2k_i k_{3-i}c_i + k_i^2\lambda_{3-i,i}), \quad (88)$$

$$H^1_{L_1} = (p_1 - c_1)\rho_1(a_1 + k_1(p_2 - p_1)) - 2\phi_1 L_1 + \lambda_{11}\rho_1(a_1 + k_1(p_2 - p_1)) + \lambda_{12}\rho_1(a_2 + k_2(p_1 - p_2)) = 0. \quad (89)$$

Therefore,

$$L_1 = \frac{\rho_1\{(p_1 - c_1 + \lambda_{11})(a_1 + k_1(p_2 - p_1)) + \lambda_{12}(a_2 + k_2(p_1 - p_2))\}}{2\phi_1}, \quad (90)$$

$$L_2 = \frac{\rho_2\{(p_2 - c_2 + \lambda_{22})(a_2 + k_2(p_1 - p_2)) + \lambda_{21}(a_1 + k_1(p_2 - p_1))\}}{2\phi_2}, \quad (91)$$

$$\dot{\lambda}_{11} = -H^1_{x_1} = (p_1 - c_1)(a_1 + k_1(p_2 - p_1)) + \lambda_{11}(a_1 + k_1(p_2 - p_1)) + \lambda_{12}(a_2 + k_2(p_1 - p_2)), \quad (92)$$

$$\dot{\lambda}_{12} = -H^1_{x_2}$$
$$= (p_1 - c_1)(a_1 + k_1(p_2 - p_1)) + \lambda_{11}(a_1 + k_1(p_2 - p_1)) + \lambda_{12}(a_2 + k_2(p_1 - p_2)). \quad (93)$$

From (92) and (93),

$$\dot{\lambda}_{11} = \dot{\lambda}_{12}, \quad (94)$$

$\lambda_{12} = \lambda_{11} + \Xi$, Ξ is constant.

Since $\lambda_{11}(T) = \frac{dS_1}{dx_1(T)} = \sigma_1$ and $\lambda_{12}(T) = \frac{dS_1}{dx_2(T)} = -\delta_1$, $S_i \equiv \sigma_i x_i(T) - \delta_i x_{3-i}(T)$, $\Xi = \lambda_{12} - \lambda_{11} = -\delta_1 - \sigma_1$. Therefore, $\lambda_{i,3-i} = \lambda_{ii} - (\sigma_i + \delta_i)$.

The solution also satisfies the sufficient condition.

From (86), $\frac{\partial^2 H^i}{\partial p_i^2} = -2k_i(\rho_1 L_1 + \rho_2 L_2 - x_1 - x_2) \leq 0$, and from (89), $\frac{\partial^2 H^i}{\partial L_i^2} = -2\phi_i \leq 0$. In addition, the constraint is jointly concave in p_i and L_i. Thus, the necessary conditions are also sufficient for optimality. Consequently, the solution is unique.

Derivation of Theorem 2

From (88),

$$p_i = \frac{1}{3k_i k_{3-i}}(2k_{3-i}a_i - k_i k_{3-i}\lambda_{3-i,3-i} + k_i k_{3-i}c_{3-i}$$
$$- 2k_i k_{3-i}\lambda_{ii} + k_i a_{3-i} + 2k_{3-i}^2\lambda_{i,3-i} + 2k_i k_{3-i}c_i + k_i^2\lambda_{3-i,i}).$$

Thus,

$$\frac{dp_i}{dt} = \dot{p}_i = \frac{1}{3k_i k_{3-i}}(-k_i k_{3-i}\dot{\lambda}_{3-i,3-i} - 2k_i k_{3-i}\dot{\lambda}_{ii} + 2k_{3-i}^2\dot{\lambda}_{i,3-i} + k_i^2\dot{\lambda}_{3-i,i}). \quad (95)$$

If $k_i = k_{3-i} = \bar{k}, \dot{p}_i = 0$ since $\dot{\lambda}_{ii} = \dot{\lambda}_{i,3-i}$, and therefore p_i is constant. (96)

Since p_i is constant, it must be that $p_i^*(t) = p_i^*(T)$.

Thus, $p_i^* = \frac{1}{3\bar{k}}(2a_i - \bar{k}\sigma_{3-i} + \bar{k}c_{3-i} - 2\bar{k}\sigma_i + a_{3-i} - 2\bar{k}\delta_i + 2\bar{k}c_i - \bar{k}\delta_{3-i})$ at $t = T$.

Therefore, $p_i^* = \frac{1}{3\bar{k}}(2a_i - \bar{k}\sigma_{3-i} + \bar{k}c_{3-i} - 2\bar{k}\sigma_i + a_{3-i} - 2\bar{k}\delta_i + 2\bar{k}c_i - \bar{k}\delta_{3-i})$ at $t \in [0, T]$.

Appendix 2

Solving the Base Model

[Stage 1]
From (64) and (65), we obtain:

$$p^*(t) = \frac{d_1 + cd_2 + d_3\tau - \lambda d_2}{2d_2}, \tag{97}$$

$$q^*(t) = \frac{\gamma}{k_1}. \tag{98}$$

Since $k_1 < k_2$, there is an incentive to make the spare parts more than needed during the first stage. Therefore, from (42), it can be assumed that $\eta = 0$ and $S(t) - F(t) > 0$, which are put into (67) to obtain $\gamma' = h$. Thus, we have:

$$\gamma(t) = ht + X_1,\ X_1\text{: constant,} \tag{99}$$

$$\theta(t) = c_1 + \frac{h}{\alpha} + X_2 e^{\alpha t},\ X_2\text{: constant.} \tag{100}$$

Putting (100) into (66), we get $\lambda' = -h - \alpha X_2 e^{\alpha t}$ and the following:

$$\lambda(t) = -ht - X_2 e^{\alpha t} + X_3,\ X_3\text{: constant.} \tag{101}$$

Therefore, combining (97) and (101), we obtain:

$$p^*(t) = \frac{1}{2}\left(\frac{d_1}{d_2} + c + \frac{d_3\tau}{d_2} - X_3\right) + \frac{h}{2}t + \frac{X_2}{2}e^{\alpha t}. \tag{102}$$

In addition, using (98) and (99), we obtain:

$$q^*(t) = \frac{h}{k_1}t + \frac{X_1}{k_1}. \tag{103}$$

Now, putting (102) into (41) and solving the differential equation, we get:

$$A(t) = \frac{1}{2}\left\{(d_1 - d_2 c + d_2 X_3 + d_3\tau)t - \frac{d_2 h}{2}t^2 - \frac{d_2 X_2}{\alpha}e^{\alpha t}\right\} + X_4,\ X_4\text{: constant.} \tag{104}$$

Since $A(0) = 0$, we have $X_4 = \frac{d_2 X_2}{2\alpha}$ and obtain:

$$A^*(t) = \frac{1}{2}\left\{(d_1 - cd_2 + d_2X_3 + d_3\tau)t - \frac{d_2h}{2}t^2 - \frac{d_2X_2}{\alpha}e^{\alpha t}\right\} + \frac{d_2X_2}{2\alpha}. \tag{105}$$

Putting (A9) into $\dot{F}(t) = \alpha\{A(t) - F(t)\}$ and solving the differential equation, we obtain:

$$\begin{aligned} F(t) &= \frac{1}{2}\left(\frac{cd_2 + d_2X_2 - d_1 - d_2X_3 - d_3\tau}{\alpha} - \frac{d_2h}{\alpha^2}\right) \\ &+ \frac{t}{2}\left(d_1 - cd_2 + d_2X_3 + d_3\tau + \frac{d_2h}{\alpha}\right) - \frac{d_2X_2e^{\alpha t}}{4\alpha} - \frac{d_2ht^2}{4} + X_5e^{-\alpha t}. \end{aligned} \tag{106}$$

Using $F(0) = 0$ and eliminating X_5, we have:

$$\begin{aligned} F^*(t) = &\frac{1}{2}\left(\frac{cd_2 + d_2X_2 - d_1 - d_2X_3 - d_3\tau}{\alpha} - \frac{d_2h}{\alpha^2}\right) + \frac{t}{2}\left(d_1 - cd_2 + d_2X_3 + d_3\tau + \frac{d_2h}{\alpha}\right) \\ &- \frac{d_2X_2e^{\alpha t}}{4\alpha} - \frac{d_2ht^2}{4} + \left[\frac{d_2X_2}{4\alpha} - \frac{1}{2}\left(\frac{cd_2 + d_2X_2 - d_1 - d_2X_3 - d_3\tau}{\alpha} - \frac{d_2h}{\alpha^2}\right)\right]e^{-\alpha t}. \end{aligned} \tag{107}$$

Since $\dot{S}(t) = q(t) = \frac{h}{k_1}t + \frac{X_1}{k_1}$, we have $S(t) = \frac{h}{2k_1}t^2 + \frac{X_1}{k_1}t + X_6$. Using $S(0) = 0$, we obtain:

$$S^*(t) = \frac{h}{2k_1}t^2 + \frac{X_1}{k_1}t. \tag{108}$$

Using a switching condition (79), we derive the relationship between X_2 and X_3. Since we produce the current product only by T, it must be that $\lambda^+(T) = 0$. Thus, (79) can be rewritten as:

$$\lambda^-(T) = 2nA(T) - m. \tag{109}$$

Using (101) and (109), we establish the relationship between X_2 and X_3 as follows:

$$X_3 = \frac{1}{(1 - nd_2T)}\left[n\left\{(d_1 - d_2c + d_3\tau)T - \frac{d_2h}{2}T^2 - \frac{d_2X_2}{\alpha}e^{\alpha T}\right\} + \frac{nd_2X_2}{\alpha} + X_2e^{\alpha T} - m + hT\right] \tag{110}$$

[Stage 2–1]

Since we consider the case $T < t^* < T + \tau$, we have $\eta = 0$, $S(t) - F(t) > 0$ for $T \leq t \leq t^*$, i.e., Stage 2–1. Using this along with (73), we obtain $\gamma' = h - \eta = h$ and

$\gamma(t) = ht + Y_1$, Y_1: constant by solving the differential equation. In addition, we have $Y_1 = X_1$ by using (77) and obtain:

$$\gamma(t) = ht + X_1 \tag{111}$$

Similarly, we obtain:

$$\theta(t) = c_2 + \frac{h}{\alpha} + Y_2 e^{\alpha t}, \tag{112}$$

$$Y_2 = (c_1 - c_2)e^{-\alpha T} + X_2. \tag{113}$$

From (105),

$$A(T) = \frac{1}{2}\left[(d_1 - d_2 c + d_2 X_3 + d_3 \tau)T - \frac{d_2 h}{2}T^2 - \frac{d_2 X_2}{\alpha}e^{\alpha T}\right] + \frac{d_2 X_2}{2\alpha}. \tag{114}$$

Since $A(t)$ is constant, we get:

$$F(t) = A(T) + Y_3 e^{-\alpha t},\ Y_3 \text{ constant.} \tag{115}$$

Since $F(t)$ must be continuous, $F(T)$ from Stage 1 must be equal to $F(T)$ from Stage 2. Using this condition, we obtain:

$$\begin{aligned} Y_3 = &\frac{1}{2}\left(\frac{cd_2 - d_1 + d_2 X_2 - d_2 X_3 - d_3 \tau}{\alpha} - \frac{d_2 h}{\alpha^2}\right)e^{\alpha T} + \frac{d_2 h}{2\alpha}e^{\alpha T}T \\ &+ \frac{d_2 X_2}{4\alpha}e^{2\alpha T} - \frac{d_2 X_2}{2\alpha}e^{\alpha T} + \left[\frac{d_2 X_2}{4\alpha} - \frac{1}{2}\left(\frac{cd_2 - d_1 + d_2 X_2 - d_2 X_3 - d_3 \tau}{\alpha} - \frac{d_2 h}{\alpha^2}\right)\right], \end{aligned} \tag{116}$$

$$\begin{aligned} F^*(t) = &\frac{1}{2}\left[(d_1 - d_2 c + d_2 X_3 + d_3 \tau)T - \frac{d_2 h}{2}T^2 - \frac{d_2 X_2}{\alpha}e^{\alpha T}\right] \\ &+ \frac{d_2 X_2}{2\alpha} + \left[\frac{1}{2}\left(\frac{cd_2 - d_1 - d_2 X_3 - d_3 \tau}{\alpha} - \frac{d_2 h}{\alpha^2}\right)e^{\alpha T} + \frac{d_2 h}{2\alpha}e^{\alpha T}T + \frac{d_2 X_2}{4\alpha}e^{2\alpha T} + \left[\frac{d_2 X_2}{4\alpha} - \frac{1}{2}\left(\frac{cd_2 - d_1 - d_2 X_3 - d_3 \tau}{\alpha} - \frac{d_2 h}{\alpha^2}\right)\right]\right]e^{-\alpha t}. \end{aligned} \tag{117}$$

From (72) and (111), $q(t) = \frac{\gamma(t)}{k_2} = \frac{h}{k_2}t + \frac{X_1}{k_2}$. Since $\dot{S} = q(t)$, solving the differential equation, we get:

$$S^*(t) = \frac{h}{2k_2}t^2 + \frac{X_1}{k_2}t + Y_4,\ Y_4 \text{ Constant.} \tag{118}$$

Similarly using the continuity of a state variable, we derive:

$$Y_4 = \left(\frac{1}{k_1} - \frac{1}{k_2}\right)\left(\frac{hT^2}{2} + X_1 T\right). \tag{119}$$

Therefore, we obtain:

$$S^*(t) = \frac{h}{2k_2}t^2 + \frac{X_1}{k_2}t + \left(\frac{1}{k_1} - \frac{1}{k_2}\right)\left(\frac{hT^2}{2} + X_1 T\right). \tag{120}$$

[Stage 2–2] For Stage 2–2, we assume $\eta > 0$ and $S(t) - F(t) = 0$. Using the necessary conditions and relevant constraints, we obtain:

$$F(t) = S(t) = A(T) + Y_5 e^{-\alpha t},\ Y_5 \text{ constant}, \tag{121}$$

$$\dot{S}(t) = \dot{A}(T) - \alpha Y_5 e^{-\alpha t} = -\alpha Y_5 e^{-\alpha t} = \mathrm{q(t)}, \tag{122}$$

Since

$$q(t) = \frac{\gamma}{k_2},\ \gamma(t) = k_2 q(t) = -\alpha k_2 Y_5 e^{-\alpha t}, \tag{123}$$

$$\dot{\gamma}(t) = \alpha^2 k_2 Y_5 e^{-\alpha t} = h - \eta(t), \tag{124}$$

$$\eta(t) = h - \alpha^2 k_2 Y_5 e^{-\alpha t}. \tag{125}$$

In order to calculate the integration constant Y_5, we assume $\eta(T+\tau) = \pi_R$, which implies that the marginal value (or salvage value) of a unit 'unused' spare part at the end of the decision time horizon is $\pi_R > 0$. Therefore, we have $\eta(T+\tau) = h - \alpha^2 k_2 Y_5 e^{-\alpha(T+\tau)} = \pi_R$ and $Y_5 = \frac{h-\pi_R}{\alpha^2 k_2} e^{\alpha(T+\tau)}$. We obtain:

$$\eta(t) = h - (h - \pi_R)e^{\alpha(T+\tau-t)}. \tag{126}$$

Using (74), we get $\theta' - \alpha\theta = -c_2\alpha - \alpha^2 k_2 Y_5 e^{-\alpha t}$, which can be solved to show:

$$\theta(t) = c_2 + \frac{k_2 Y_5}{2}\alpha e^{-\alpha t} + Y_6 e^{\alpha t},\ Y_6 \text{ constant}. \tag{127}$$

Using the continuity condition of $\theta(t)$ at t^*, we have:

$$\theta(t^*) = c_2 + \frac{k_2 Y_5}{2}\alpha e^{-\alpha t^*} + Y_6 e^{\alpha t^*} = \frac{h}{\alpha} + c_2 + X_2 e^{\alpha t^*}. \tag{128}$$

Solving (128), we get:

$$Y_6 = \frac{h}{\alpha} e^{-\alpha t^*} + Y_2 - \frac{\alpha k_2 Y_5}{2} e^{-2\alpha t^*}. \tag{129}$$

In order to derive the integration constant X_1, we use the continuity of $\gamma(t)$ at t^*. Since $\gamma(t^*) = ht^* + X_1 = -\alpha k_2 Y_5 e^{-\alpha t^*}$ from (99) and (123), and we already know $Y_5 = \frac{h-\pi_R}{\alpha^2 k_2} e^{\alpha(T+\tau)}$, we obtain:

$$X_1 = -ht^* - \frac{(h - \pi_R)}{\alpha} e^{\alpha(T+\tau-t^*)}. \tag{130}$$

In order to calculate t^*, we use the continuity of $S(t)$.

That is, we know that $S(t) = \frac{h}{2k_2} t^2 + \frac{X_1}{k_2} t + \left(\frac{1}{k_1} - \frac{1}{k_2}\right)\left(\frac{hT^2}{2} + X_1 T\right)$ for $t \le t^*$ and $S(t) = A(T) + \frac{(h-\pi_R)}{\alpha^2 k_2} e^{\alpha(T+\tau-t)}$ for $t^* \le t$, and t^* must satisfy (131):

$$\frac{h}{2k_2} {t^*}^2 + \frac{X_1}{k_2} t^* + \left(\frac{1}{k_1} - \frac{1}{k_2}\right)\left(\frac{hT^2}{2} + X_1 T\right) = A(T) + \frac{(h - \pi_R)}{\alpha^2 k_2} e^{\alpha(T+\tau-t^*)}. \tag{131}$$

We derive X_2 and X_3, using the dynamic pattern of $F(t)$. That is, since (115) and (121) must be the same, it holds $Y_3 = Y_5$, which implies (132):

$$\begin{aligned}
&\frac{1}{2}\left(\frac{cd_2 - d_1 + d_2 X_2 - d_2 X_3 - d_3 \tau}{\alpha} - \frac{d_2 h}{\alpha^2}\right) e^{\alpha T} + \frac{d_2 h}{2\alpha} e^{\alpha T} T + \frac{d_2 X_2}{4\alpha} e^{2\alpha T} - \frac{d_2 X_2}{2\alpha} e^{\alpha T} \\
&+ \left[\frac{d_2 X_2}{4\alpha} - \frac{1}{2}\left(\frac{cd_2 - d_1 + d_2 X_2 - d_2 X_3 - d_3 \tau}{\alpha} - \frac{d_2 h}{\alpha^2}\right)\right] = \frac{h - \pi_R}{\alpha^2 k_2} e^{\alpha(T+\tau)}.
\end{aligned} \tag{132}$$

Therefore, using (110) and (132), we now can derive X_2 and X_3.

How to Determine the Optimal T

The switching condition (76) can be rewritten as (133):

$$\begin{aligned}
&- d_2 p(T)^2 + (d_1 + cd_2) p(T) - cd_1 - \frac{k_1}{2} q(T)^2 - c_1\{\alpha(A(T) - F(T))\} + \frac{\partial R(T)}{\partial T} \\
&= -\frac{k_2}{2} q(T)^2 - c_2\{\alpha(A(T) - F(T))\}.
\end{aligned} \tag{133}$$

Since $R(T) = mA(T) - nA(T)^2$, we have (134):

$$\frac{\partial R(T)}{\partial T} = m\frac{\partial A(T)}{\partial T} - 2nA(T)\frac{\partial A(T)}{\partial T}. \tag{134}$$

Now, since $A(T) = \frac{1}{2}\left[(d_1 - cd_2 + d_2X_3)T - \frac{d_2h}{2}T^2 - \frac{d_2}{\alpha}X_2e^{\alpha T}\right] + \frac{d_2}{\alpha}X_2$, and X_2 and X_3 are functions of T, we have:

$$\frac{\partial A(T)}{\partial T} = \frac{1}{2}\left(d_2\frac{\partial X_3}{\partial T} + d_1 - cd_2 + d_2X_3\right) - \frac{d_2h}{2}T - \frac{d_2}{2\alpha}\left(\frac{\partial X_2}{\partial T} + X_2\alpha e^{\alpha T}\right) + \frac{d_2}{\alpha}\frac{\partial X_2}{\partial T} \tag{135}$$

Plugging (135) into (132) and (133) and rearranging appropriately, we obtain:

$$\begin{aligned} &-d_2p(T)^2 + (d_1 + cd_2)p(T) - cd_1 - \frac{q(T)^2}{2}(k_1 - k_2) \\ &- \alpha(A(T) - F(T))(c_1 - c_2) + \frac{\partial R(T)}{\partial T} \\ &= 0, \end{aligned} \tag{136}$$

$$p(T) = \frac{1}{2}\left(\frac{d_1}{d_2} + c + \frac{d_3\tau}{d_2} - X_3\right) + \frac{h}{2}T + \frac{X_2}{2}e^{\alpha t}, \tag{137}$$

$$q(T) = \frac{h}{k_1}T + \frac{X_1}{k_1}, \tag{138}$$

$$\begin{aligned} \frac{\partial R(T)}{\partial T} = {} & m\frac{1}{2}\left\{\left(d_2\frac{\partial X_3}{\partial T} + d_1 - d_2c + d_2X_3\right) - \frac{d_2h}{2}T - \frac{d_2}{2\alpha}\left(\frac{\partial X_2}{\partial T} + X_2\alpha e^{\alpha T}\right) + \frac{d_2}{\alpha}\frac{\partial X_2}{\partial T}\right\} \\ & - 2n\left\{\frac{1}{2}\left[(d_1 - d_2c + d_2X_3)T - \frac{d_2h}{2}T^2 - \frac{d_2}{\alpha}X_2e^{\alpha T}\right] + \frac{d_2}{\alpha}X_2\right\} \\ & \times\left\{\frac{1}{2}\left(d_2\frac{\partial X_3}{\partial T} + d_1 - d_2c + d_2X_3\right) - \frac{d_2h}{2}T - \frac{d_2}{2\alpha}\left(\frac{\partial X_2}{\partial T} + X_2\alpha e^{\alpha T}\right) + \frac{d_2}{\alpha}\frac{\partial X_2}{\partial T}\right\}. \end{aligned} \tag{139}$$

The T that satisfies (136) is the optimal T. In order to solve (136), we need to know t^*, which can be derived from (131). Therefore, by solving (131) and (136) simultaneously, we can obtain optimal t^* and T at the same time.

Chapter 5
Sustainable Supply Chain Management

Abstract Sustainability is the spirit of the time at the moment. As such, sustainable supply chain management is a very important and relevant topic and deserves to be in the concluding chapter. First, I premise that in addition to firms, there are two key players in the economy, the government and consumers, who should exert an influence on pollution reduction. I explore whether these two are substitutes or complements in improving environment. Then, I embark on analyzing the issue from the value chain perspective by looking into the role of consumer awareness in light of supply chain coordination strategy. I show why and how consumer awareness can make a huge impact on environmental sustainability.

Keywords Sustainability · Sustainable supply chain management · Government · Consumer awareness · Pollution · Value chain perspective · Environmental sustainability

Key Learning

- *How do the government and the market affect firm's pollution abatement efforts, i.e., firm's efforts to reduce its pollution emission?*
  - The government penalty and the consumer's awareness are substitutes either when the market size is relatively large or when the market is relatively small, but the government penalty is relatively heavy.
  - The two factors are complements when the market size is relatively small and the government penalty is relatively light.
- *To understand the dynamics of the manufacturer's effort to reduce pollution in a supply chain consisting of manufacturer, retailer, and consumers, we analyze four cases according to consumer awareness of the pollution's harmful effect, i.e., environmentally aware versus ignorant, and supply chain coordination, i.e., competitive versus cooperative.*

The original version of this chapter was revised. The erratum to this chapter is available at 10.1007/978-981-10-3599-9_6

B. Kim, *Optimal Control Applications for Operations Strategy*,
DOI 10.1007/978-981-10-3599-9_5

- The supply chain coordination strategy becomes irrelevant to reducing the pollution, if the consumers are not environmentally aware or sensitive enough.
- Transfer price and potential market size are the important factors to determine each case's relative effectiveness.

Nowadays, sustainability is a key issue in supply chain management. Sustainable supply chain management requires the firm to approach its value chain management from diverse perspectives of its internal and external stakeholders. Consistent with the objectives of sustainable supply chain management, in this chapter, we discourse issues on pollution reduction from the perspective of the various stakeholders sharing the same supply chain, focusing on the environmental sustainability, i.e., how to reduce the pollution. In particular, we examine the dynamic interaction between value chain partners in affecting the pollution reduction effort. We first examine how the government and the market affect firm's pollution abatement efforts, i.e., firm's efforts to reduce its pollution emission. The way for the government to control firm's pollution is to impose penalty, whereas the consumers (the market) make their purchasing decision by taking into account the pollution, i.e., the demand is affected by the stock of pollution. In effect, we consider two forces, government penalty and consumer's sensitivity to pollution, as the primary factors to control firm's pollution and analyze their interaction in relation to the firm's pollution reduction efforts. Then, we try to understand the dynamics of the manufacturer's effort to reduce pollution in a supply chain consisting of manufacturer, retailer, and consumers, by analyzing four cases according to consumer awareness of the pollution's harmful effect, i.e., environmentally aware versus ignorant, and supply chain coordination, i.e., competitive versus cooperative.

1 Role of Government and Consumers in Pollution Reduction[1]

Environmental sustainability has become a critical issue both economically and managerially. That is, it is an important issue not only for the economy, but also for the firm, since the vast majority of pollution is emitted during the firm's production process. There are stakeholders, who are concerned about the pollution in the environment. For instance, to curb the economic disutility, the government might impose a penalty on the firm for its emitting pollution. Pollution could also directly affect the utility of the consumer: It might reduce the consumer's utility

[1]This section is a refined version of Kim, B. and J.E. Sim (2015) 'Impacts of Government and Market on Firm's Efforts to Reduce Pollution'. *Cogent Economics and Finance, 3* (1), DOI:10.1080/23322039.2015.1062634.

(Agrawal et al. 2012). If the consumer dislikes and is sensitive to the pollution, she would adjust her demand for the product by the firm, which emits pollution. That is, the way the consumer penalizes the firm for its pollution emission is to reduce her demand for the firm's product. In this chapter, we model the firm's investment in reducing the pollution, which is affected by two factors, the government penalty and the consumer's sensitivity to the firm's pollution. In examining the dynamics of firm's environmental efforts, differential games and optimal control theory models have been widely developed. For instance, El Ouardighi et al. (2014) examined the optimal control problem of production and emissions reductions, focusing on the absorption capacity of the environment.

1.1 Optimal Control Theory Model and Analysis Outcomes

Our analysis context is described in Fig. 1, where there are three economic entities, the government, consumer, and the firm. The firm emits pollution while producing its product. In order to minimize the pollution, the government imposes a penalty on the firm for its pollution emission. Since the pollution reduces the consumer utility, the consumer also wants to penalize the firm for its pollution by reducing her demand for the firm's product. In developing the firm's optimal control theory model, we define the variables and parameters as in Table 1.

The firm's objective function writes:

$$Maximize\, J = \int_0^\infty e^{-rt}\left[(p-c)D - c_1\left(D-\overline{U}\right)^2 - ev^2 - fy^2\right]\mathrm{d}t \tag{1}$$

In (1), $(p-c)D$, where $p \geq 0$ is the total net profit for the firm, where $(p-c)$ is the unit profit, sales price minus the unit production cost, and $D = \alpha - \beta p - \gamma y$ is

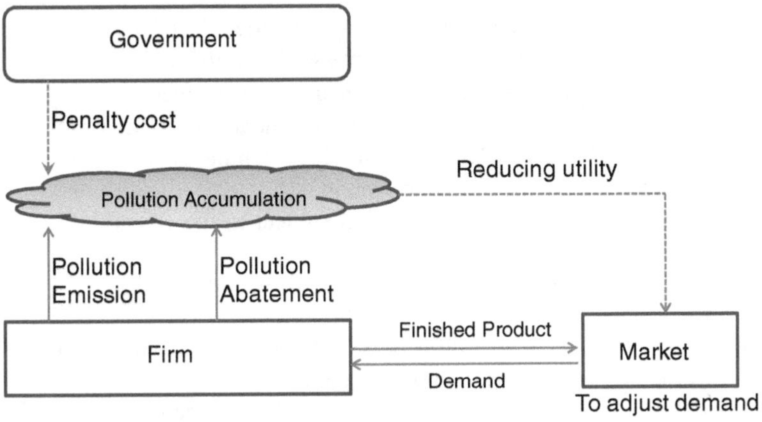

Fig. 1 General context of sustainable value chain

Table 1 Definitions of variables and parameters

$y(t)$	Cumulative pollution at t
$v(t)$	Firm's effort to reduce the emission of pollutants at t
f	Cost parameter associated with government's penalty on the cumulative pollution
e	Cost parameter associated with firm's pollution abatement effort
$\overline{U}$	Firm's plant capacity
l	Unit pollutant emission per manufacturing capacity
$p(t)$	Sales price at t
c	Unit production cost of the product
c_1	Cost parameter associated with the deviation from the manufacturing capacity $\overline{U}$
$D(t)$	Demand at t; $D = \alpha - \beta p - \gamma y$
α	Potential market size
β	Coefficient in the demand function associated with the sales price
γ	Coefficient in the demand function associated with the cumulative pollution
δ	Decay rate of the cumulative pollution
r	Discount rate
y_{LR}	Long-run equilibrium of cumulative pollution
v_{LR}	Long-run equilibrium of firm's pollution abatement effort
p_{LR}	Long-run equilibrium of the sales price

the demand function, i.e., the consumer's demand for the product is a function of the sales price, p, and the pollution stock, y: the higher the sales price and the larger the pollution stock is, the smaller the demand will be. In addition, a cost occurs in a quadratic pattern as the total production amount deviates from the firm's effective capacity, $\overline{U}$: the more the production amount deviates from the effective capacity, the larger the quadratic cost, i.e., $c_1\left(D-\overline{U}\right)^2$. While manufacturing the product, the firm emits pollutants harmful to the environment: y is the stock of pollution accumulated by t. The government imposes a penalty on the pollution stock, i.e., fy^2. This quadratic form of cost related to the pollution stock is often utilized in the environmental studies (Bertinelli et al. 2014). In order to reduce the government penalty, the firm makes an effort to cut its emission of pollutants. The effort level is denoted as v, and an associated cost is incurred in a quadratic way like ev^2, implying the increasing marginal cost of the abatement activity as widely assumed in the literature (e.g., Ni et al. 2010; Liu et al. 2012; Chung et al. 2013). Finally, the firm's profit is discounted with the rate r as in e^{-rt}.

The firm maximizes its objective function subject to the constraint:

$$\dot{y} = \overline{U}(l - v) - \delta y,\ y(0) = y_0 > 0,\ \text{where}\ 0 \leq v < l.$$

If the firm does not make any effort to reduce the pollution, it emits pollution as much as $\overline{U}l$ at t, i.e., the amount of pollution emission is proportional to the firm's capacity (Laplante and Rilstone 1996; Grant et al. 2002): one unit of capacity emits l units of pollution. If the firm's effort level to reduce the pollution is v, one unit of

Table 2 Summary of long-term equilibrium solutions

Variable	Long-run equilibrium
y_{LR}	$\frac{4\beta el\overline{U}(1+\beta c_1)(r+\delta)-\gamma\overline{U}^2\left(\alpha-\beta c+2\beta c_1\overline{U}\right)}{4\beta(1+\beta c_1)\left[f\overline{U}^2+e\delta(r+\delta)\right]-\gamma^2\overline{U}^2}$
v_{LR}	$l-\frac{\delta}{\overline{U}}\cdot y_{LR}$
p_{LR}	$\frac{\alpha(1+2\beta c_1)-2\beta c_1\overline{U}+\beta c}{2\beta(1+\beta c_1)}-\frac{\gamma(1+2\beta c_1)}{2\beta(1+\beta c_1)}\cdot y_{LR}$

capacity emits only $(l-v)$ units of pollution. The pollution stock y decays naturally by δy at t. Now we have the dynamic evolution of pollution stock as $\dot{y}=\overline{U}(l-v)-\delta y$. The resulting optimization problem of the firm is written as follows:

Base Model

$$Maximize\, J=\int_0^\infty e^{-rt}\left[(p-c)(\alpha-\beta p-\gamma y)-c_1\left(\alpha-\beta p-\gamma y-\overline{U}\right)^2-ev^2-fy^2\right]\mathrm{d}t$$

$$Subject\ to\ \dot{y}=\overline{U}(l-v)-\delta y$$
$$y(0)=y_0>0,\quad \text{where}\quad 0\leq v<l\quad \text{and}\quad p\geq 0$$

After solving the optimal control theory model, we summarize the analysis results for the long-term equilibrium in Table 2.

1.2 Theorems

Since we are interested in the long-term dynamics of pollution reduction efforts at the firm level, we postulate theorems to characterize the long-term equilibrium relationship between government penalty and consumer awareness in enhancing the firm's activity for pollution abatement. The first theorem is concerned with the effect of government penalty (*f*) on the firm's effort to mitigate the pollution (v_{LR}) and the pollution stock (y_{LR}): The government penalty imposed on the firm's pollution increases the firm's long-term effort to reduce the pollution, which in turn decreases the long-term pollution stock.

Theorem 1 *It holds that,* $\frac{\partial y_{LR}}{\partial f}<0, \frac{\partial^2 y_{LR}}{\partial f^2}>0, \frac{\partial v_{LR}}{\partial f}>0$ *and* $\frac{\partial^2 v_{LR}}{\partial f^2}<0$

Proof See Appendix "Proof of Theorem 1".

Theorem 2 postulates the interaction between government penalty and consumer awareness, i.e., whether the two forces are complements or substitutes. The interaction relationship depends on the market size and the initial magnitude of the government penalty.

Theorem 2 *For any feasible* (γ, f):

(i) *if* $G \leq 0$, $\frac{\partial^2 y_{LR}}{\partial f \partial \gamma} \geq 0$, *and* $\frac{\partial^2 v_{LR}}{\partial f \partial \gamma} \leq 0$ *for all* f
(ii) *if* $G > 0$, $\frac{\partial^2 y_{LR}}{\partial f \partial \gamma} \geq 0$, *and* $\frac{\partial^2 v_{LR}}{\partial f \partial \gamma} \leq 0$ *for* $f \geq h(\gamma)$
(iii) *if* $G > 0$, $\frac{\partial^2 y_{LR}}{\partial f \partial \gamma} < 0$, *and* $\frac{\partial^2 v_{LR}}{\partial f \partial \gamma} > 0$ *for* $f < h(\gamma)$, *where*

$$A = 3\overline{U}^4\left(\alpha - \beta c + 2\beta c_1 \overline{U}\right) > 0,$$
$$B = 16\beta el\overline{U}^3(1 + \beta c_1)(r + \delta) > 0,$$
$$C = 4\beta \overline{U}^4(1 + \beta c_1)\left(\alpha - \beta c + 2\beta c_1 \overline{U}\right) > 0,$$
$$D = 4\beta e\delta \overline{U}^2(r + \delta)(1 + \beta c_1)\left(\alpha - \beta c + 2\beta c_1 \overline{U}\right) > 0,$$
$$G = B^2 - 4AD = 16\beta e\overline{U}^6(r + \delta)(1 + \beta c_1)\left[16\beta el^2(1 + \beta c_1)(r + \delta) - 3\delta\left(\alpha - \beta c + 2\beta c_1 \overline{U}\right)^2\right]$$

$$G \leq 0, \text{ if } \alpha \geq \tilde{\alpha} = \sqrt{\frac{16\beta el^2(1 + \beta c_1)(r + \delta)}{3\delta}} + \beta c - 2\beta c_1 \overline{U}; \; G > 0, \text{ if } \alpha < \tilde{\alpha}$$
$$h(\gamma) = -\frac{A}{C}\gamma^2 + \frac{B}{C}\gamma - \frac{D}{C} = -\frac{A}{C}\left(\gamma - \frac{B}{2A}\right)^2 + \frac{B^2}{4AC} - \frac{D}{C}$$

Proof See Appendix "Proof of Theorem 2".

We summarize the implications of Theorem 2 as in Fig. 2. First, we need to consider two separate cases: one where the market is relatively large (i.e., $\alpha \geq \tilde{\alpha}$)

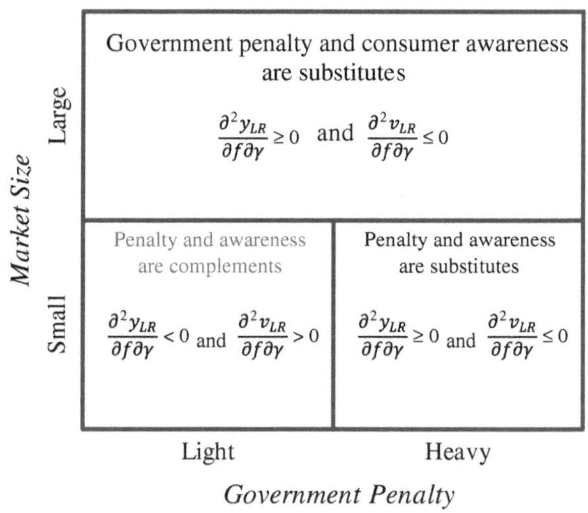

Fig. 2 Summary results

Table 3 Parameter values for the numerical analysis

r	c	c_1	e	f	l	δ	α	β	γ	$\overline{U}$	y_0
0.004	20	10^{-5}	10^9	0.01	0.01	0.1	200,000	2,000	0.1	100,000	0

Table 4 Long-term equilibrium for base case

Variable	Long-term equilibrium
Cumulative pollution (y)	760.07
Pollution abatement effort (v)	0.00924
Sales price (p)	59.78
Market demand per period	80,355
Firm's profit for the entire period	7.75595×10^8

and the other where the market is relatively small (i.e., $\alpha < \tilde{\alpha}$). The other criterion is the current level of government penalty in relation to the consumer awareness, i.e., relatively heavy ($f \geq h(\gamma)$) or relatively light ($f < h(\gamma)$). Using the two criteria, we can recapitulate the analysis results. When the market size is relatively large, government penalty and consumer awareness are substitutes, implying that for a large market, it is better to utilize either government penalty or consumer awareness, but not both simultaneously. Such a substitute relationship remains the same when the market is relatively small but the current government penalty is heavy. Finally, government penalty and consumer awareness are complements when the market size is small and the government penalty is light. It implies that the two factors are complementing each other, e.g., the government penalty enhances the positive effect of consumer awareness on the firm's effort to reduce the pollution and vice versa.

To visualize the analysis outcomes, we conduct a numerical analysis with the parameter values in Table 3, based on the smartphone manufacturing industry in Korea. Table 4 shows the long-term equilibrium values of the numerical analysis.

In Fig. 3, we show the relationship between government penalty and consumer awareness. Since $h(\gamma)$ is a concave function with a negative f-intercept $\left(= -\frac{D}{C}\right)$ and the two roots satisfying $h(\gamma) = 0$ are strictly positive when they are real, general patterns of Theorem 2 can be illustrated in Fig. 3a, b.

In Fig. 3a, the function $h(\gamma)$ lies under the γ-axis, where $f = 0$. Therefore, $G \leq 0$, $\frac{\partial^2 y_{LR}}{\partial f \partial \gamma} \geq 0$, and $\frac{\partial^2 v_{LR}}{\partial f \partial \gamma} \leq 0$ hold for all f. In Fig. 3b, however, a feasible (γ, f) pair falls into one of two regions, i.e., $f \geq h(\gamma)$ or $f < h(\gamma)$: for $f < h(\gamma)$, $\frac{\partial^2 y_{LR}}{\partial f \partial \gamma} < 0$, and $\frac{\partial^2 v_{LR}}{\partial f \partial \gamma} > 0$ hold, but for $f \geq h(\gamma)$, $\frac{\partial^2 y_{LR}}{\partial f \partial \gamma} \geq 0$ and $\frac{\partial^2 v_{LR}}{\partial f \partial \gamma} \leq 0$.

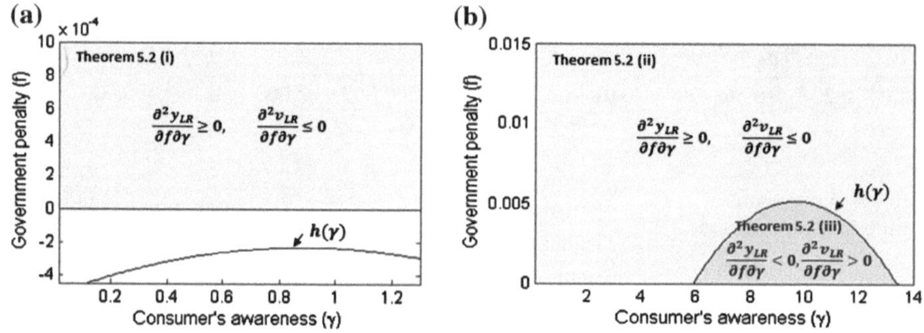

Fig. 3 Interaction between government penalty and consumer awareness. *Note* If $f \geq h(\gamma)$, $\frac{\partial^2 y_{LR}}{\partial f \partial \gamma} \geq 0$ and $\frac{\partial^2 v_{LR}}{\partial f \partial \gamma} \leq 0$ hold; if $f < h(\gamma)$, $\frac{\partial^2 y_{LR}}{\partial f \partial \gamma} < 0$ and $\frac{\partial^2 v_{LR}}{\partial f \partial \gamma} > 0$ hold. In **a**, $e = 5 \times 10^8, l = 0.02, G < 0$ and $\alpha = 100{,}000 > \tilde{\alpha} = 83{,}571$. In **b**, $e = 2.8 \times 10^{10}$, $G > 0$ and $\alpha = 200{,}000 < \tilde{\alpha} = 213{,}996$. All the other parameters are the same as in the base case

1.3 Discussion and Conclusion

We discuss economic as well as managerial implications of the analysis results. In most cases, the government penalty and the consumer awareness are substitutes. As such, implementing either one of the measures is sufficient. For instance, if the consumers are already very sensitive to the pollution, the government does not have to impose an extra penalty on the firm's pollution emission. Similarly, if the government penalty on the firm's pollution is already heavy, the consumer awareness is of little use. Another intriguing implication is that if it is difficult to educate consumers to become more sensitive to pollution, then it is better for the government to impose heavy penalty on the firm's pollution emission rather than to spend resources on educating the consumers. On the contrary, if the consumers are very sophisticated and well educated so as to be aware of pollution problems, then the government is better not to intervene by imposing an extra penalty: such an excessive intervention will reduce the effectiveness of the already high consumer awareness.

Theorem 2 and the numerical example in Fig. 3b, however, indicate that it is possible for the two factors to be complements in a relatively small pocket of the feasible region, where the market size is relatively small and the government penalty is relatively light (in relation to the consumer awareness). That is, when the market potential is relatively small and the government penalty is relatively light (i.e., $f < h(\gamma)$, where $h(\gamma) = -\frac{A}{C}\gamma^2 + \frac{B}{C}\gamma - \frac{D}{C}$), the government penalty enhances the effectiveness of the consumer awareness on the pollution reduction and vice versa. In this situation, the two measures should be implemented together to maximize the firm's effort to reduce the pollution. Note that when the government penalty becomes larger than the threshold level (i.e., $f \geq h(\gamma)$), the relationship between government penalty and consumer awareness changes to substituting.

The economic implication is clear. In most cases, the policy maker can have more latitude to choose an appropriate measure, either imposing penalty or enhancing consumer awareness in order to push the firm to make more efforts to reduce the pollution. In this situation, she does not have to implement two measures simultaneously, since doing so reduces the policy effectiveness. When the market is relatively small and the government penalty is not yet heavy, the situation can be delicate: the policy maker should skillfully implement both measures simultaneously in order to maximize the policy effectiveness.

2 Supply Chain Coordination and Consumer Awareness for Pollution Reduction[2]

Sustainable supply chain management is an immensely important issue both managerially and economically, as the environmental concerns are increasingly becoming central to global economic as well as political arenas (see Appendix 3). Srivastava (2007) defined green supply chain management by postulating 'Adding the "green" component to supply chain management involves addressing the influence and relationships between supply chain management and the natural environment.' In this section, we investigate whether and how supply chain coordination and consumer awareness affect the pollution accumulation in a supply chain. In the literature, there are two modes of supply chain coordination, competitive and cooperative. In a competitive supply chain, firms sharing the same supply chain make decisions competitively, as if they were competitors. On the contrary, in a cooperative supply chain, they make decisions for their common goals, e.g., joint-profit maximization. We also note that there are two driving forces behind the growing importance of pollution reduction in supply chain management. On the one hand, there is a government regulation, which forces the business to reduce its emission of pollutant. On the other hand, there is a collective power of consumers, whose purchasing decision can send a strong signal to the business. In this context, we define important questions to ask, 'Which one, supply chain coordination or consumer awareness, is more conducive to reducing the pollution? Is there any relationship between the two in minimizing the pollution emission? Which one, government regulation or consumer awareness, is more powerful in curbing the pollution?' In order to answer these questions, we develop four differential game models, using two dimensions, consumer awareness (aware versus ignorant consumers) and supply chain coordination (competitive versus cooperative).

---

[2]This section is a refined version of Kim, B. and J.E. Sim (2016). 'Supply Chain Coordination and Consumer Awareness for Pollution Reduction.' *Sustainability 2016*, 8, 365.

2.1 Differential Game Models and Analysis Outcome

The setting of our model can be described as in Fig. 4. The supply chain consists of three primary players—manufacturer, retailer, and consumer. The manufacturer produces and sells its product to the retailer, who in turn sells the product to the end consumer. It is the manufacturer that emits pollution during the production process, which causes the environment to deteriorate. The government imposes penalty on the manufacturer for its pollution emission.

In this context, we focus on two dimensions, i.e., supply chain coordination and consumer awareness. There are two different supply chain coordination modes, competitive and cooperative. Under the competitive supply chain coordination, each of the two supply chain participants, i.e., manufacturer and retailer, makes its own decision so as to maximize its own profit. On the contrary, under the cooperative supply chain coordination, the two participants are making a decision as if both belong to the same decision-making entity, i.e., they have one objective function combining their profits together. In addition, we consider two types of consumer, one who is aware of and sensitive to the pollution emitted by the manufacturer and the other who is ignorant of and insensitive to it. If the consumer is aware of and sensitive to the pollution emitted by the manufacturer, she will take into account the pollution level when making a purchasing decision. That is, the consumer's demand function is affected by the level of the pollution stock. Using these two dimensions, we develop and analyze four models (Fig. 5), ignorant consumer and competitive supply chain (Model 1), aware consumer and competitive supply chain (Model 2), aware consumer and cooperative supply chain (Model 3), and finally, ignorant consumer and cooperative supply chain (Model 4). In addition, without loss of generality, we assume that there is little information or time gap between supply chain players, e.g., the retailer places an order, which is

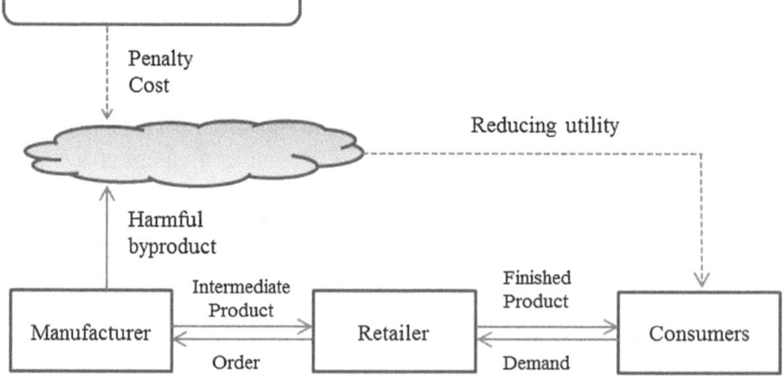

Fig. 4 General context of sustainable value chain

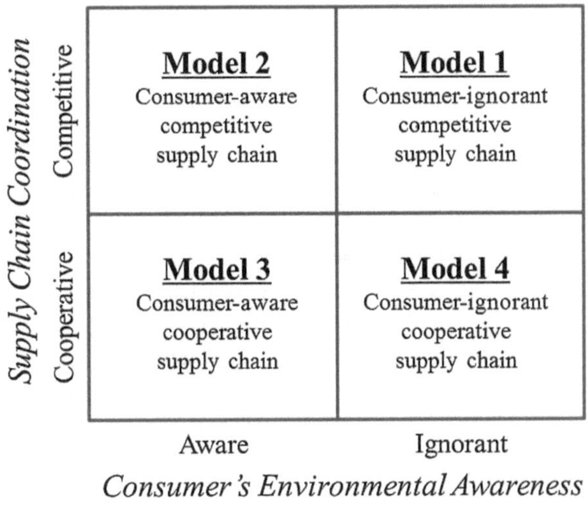

Fig. 5 Differential game models

exactly the amount demanded by the consumers, the manufacturer produces the exact amount ordered by the retailer, and all of these ordering, producing, and delivering occur within the same sales cycle.

The major variables and parameters in the models are described in Table 5. As the base case, we consider Model 1, where consumers are ignorant of the environmental issues and the supply chain is competitive. First, the retailer's objective function writes:

$$\begin{aligned} \mathit{Maximize}\, J^r &= \int_0^\infty e^{-rt}\left[(p_2 - p_1)D - c_2 D^2\right]\mathrm{d}t \\ &= \int_0^\infty e^{-rt}\left[(p_2 - p_1)(\alpha - \beta p_2) - c_2(\alpha - \beta p_2)^2\right]\mathrm{d}t \end{aligned} \tag{2}$$

In (2), $(p_2 - p_1)(\alpha - \beta p_2)$, $p_2 \geq 0$, is the total net profit for the retailer, where $(p_2 - p_1)$ is the unit profit, sales price minus transfer price paid to the manufacturer, and $(\alpha - \beta p_2)$ is the demand function, i.e., the consumer's demand for the retailer's product is a function of the sales price, p_2, charged by the retailer. A quadratic cost function is assumed so that $c_2(\alpha - \beta p_2)^2$ is the total cost required for the retailer to process $(\alpha - \beta p_2)$ units, to finalize or refine the semi-finished products from the manufacturer to sell them to the end consumers. A quadratic cost function is widely used in the literature to represent the effect of increasing marginal cost. Similar argument can be made to retailer's processing cost, for instance, due to the need of having more expensive resources for processing as the volume increases (Eliashberg and Steinberg 1987; El Ouardighi et al. 2013).

Table 5 Definition of variables and parameters

Notation	Definition
$y(t)$	Cumulative pollution at time t
$v(t)$	Effort level of the manufacturer to reduce the pollution emission
f	Cost parameter associated with the government penalty on the cumulative pollution
e	Cost parameter associated with the manufacturer's pollution abatement effort
$\overline{U}$	Manufacturer's plant capacity
l	Pollution emission per unit manufacturing capacity
$p_2(t)$	Retail price charged by the retailer at time t
$p_1(t)$	Transfer price paid to the manufacturer at time t
c	Unit production cost of the manufacturer
c_1	Cost parameter associated with the deviation from the manufacturing capacity $\overline{U}$
c_2	Cost parameter associated with the retailer's processing the product
$D(t)$	Demand for the product at time t, i.e., $D(t) = \alpha - \beta p_2(t)$ or $D(t) = \alpha - \beta p_2(t) - \gamma y(t)$
α	Potential market size
β	Coefficient in the demand function associated with the sales price $p_2(t)$
γ	Coefficient in the demand function associated with the cumulative pollution $y(t)$
δ	Decay rate of the cumulative pollution
r	Discount rate
y^i_{LR}	Long-run equilibrium of cumulative pollution in model i, $i = \text{I, II, III, IV}$
v^i_{LR}	Long-run equilibrium of the manufacturer's pollution abatement effort in model i
p^i_{2LR}	Long-run equilibrium of the sales price in model i
J, J^m, J^r	Objective function (net profit) of the whole supply chain, manufacturer, or retailer for $t \in [0, \infty)$, respectively

Similarly, the manufacturer's objective function writes:

$$\begin{aligned} Maximize\, J^m &= \int_0^\infty e^{-rt}\left[(p_1 - c)D - c_1\left(D - \overline{U}\right)^2 - ev^2 - fy^2\right]\mathrm{d}t \\ &= \int_0^\infty e^{-rt}\left[(p_1 - c)(\alpha - \beta p_2) - c_1\left(\alpha - \beta p_2 - \overline{U}\right)^2 - ev^2 - fy^2\right]\mathrm{d}t \end{aligned} \tag{3}$$

The manufacturer's total net profit is $(p_1 - c)(\alpha - \beta p_2)$, where c denotes the unit production cost. In addition to the production cost, a cost incurs in a quadratic pattern as the production amount deviates from the manufacturer's effective capacity, $\overline{U}$: the more the production amount deviates from the effective capacity, the larger the quadratic cost will be, i.e., $c_1\left(\alpha - \beta p_2 - \overline{U}\right)^2$. While producing the product, the manufacturer emits pollutants harmful to the environment: y is the stock of pollution accumulated by t. The government imposes a penalty on the pollution stock, i.e., fy^2. The assumption of increasing convex cost on the firm's pollution is often used. It also implies that pressures from a regulator increase more than proportionally with pollution amount. Weil (1997) showed that larger firms, typically generating more pollution, not only have higher probabilities of being subject to regulator's inspections, but also pay substantially higher fines per violation than small firms do. Similarly, Gray and Shadbegian (2004) showed that regulators conduct more pollution-related inspections and enforcements on larger plants. In order to reduce the government's penalty, the manufacturer makes an effort to cut its emission of pollutants. The effort level is denoted as v, and an associated cost incurs in a quadratic way like ev^2.

Both the manufacturer and the retailer maximize their objectives subject to the common constraint[3]:

$$\dot{y} = \overline{U}(l - v) - \delta y,\ y(0) = y_0 > 0, \quad \text{where} \quad 0 \le v < l. \tag{4}$$

If the manufacturer does not make any effort to reduce the pollution, it emits pollution as much as $\overline{U}l$ at t, i.e., the amount of pollution emission is proportional to the manufacturer's capacity: One unit of capacity emits l units of pollution. If the manufacturer's effort level to reduce the pollution is v, one unit of capacity emits only $(l - v)$ units of pollution. The pollution stock y decays naturally by δy at t: that is, the nature has a certain level of power to decompose and neutralize the pollutant, depending on its carrying capacity (Bertinelli et al. 2014; Tidball and Zaccour 2009). Now we have the dynamic evolution of pollution stock as $\dot{y} = \overline{U}(l - v) - \delta y$. We recapitulate the differential games for Model 1 as follows:

Model 1
Retailer's objective writes:

$$\mathit{Maximize}\, J^r = \int_0^\infty e^{-rt}\left[(p_2 - p_1)(\alpha - \beta p_2) - c_2(\alpha - \beta p_2)^2\right] \mathrm{d}t \tag{5}$$

Manufacturer's objective writes:

$$\mathit{Maximize}\, J^m = \int_0^\infty e^{-rt}\left[(p_1 - c)(\alpha - \beta p_2) - c_1\left(\alpha - \beta p_2 - \overline{U}\right)^2 - ev^2 - fy^2\right] \mathrm{d}t \tag{6}$$

$$\mathit{Subject\ to}\, \dot{y} = \overline{U}(l - v) - \delta y \tag{7}$$

$$y(0) = y_0 > 0, \quad \text{where} \quad 0 \le v < l \quad \text{and} \quad p_2 \ge 0 \tag{8}$$

[3]Note that the common constraint (4), known as a state equation, between players is a distinguishing feature of a differential game and it implies that both players have the possibility of being influenced by the state of the system (i.e., pollution stock in this model) over the planning horizon in determining their actions.

Now let us consider Model 2, which is different from Model 1 in one aspect: Model 2 assumes that the consumers are aware of and sensitive to the pollution emitted by the manufacturer, whereas Model 1 assumes that the consumers are ignorant and insensitive. How can we model the consumer's awareness? In order to incorporate the consumer's awareness of the pollution, we change the demand function so that it is now a function of not only the sales price, but also the pollution stock, i.e., $\alpha - \beta p_2 - \gamma y$, where γ is the demand function's coefficient associated with the pollution stock. We provide the decision problems as follows:

Model 2
Retailer's objective writes:

$$Maximize\, J^r = \int_0^\infty e^{-rt}\left[(p_2 - p_1)(\alpha - \beta p_2 - \gamma y) - c_2(\alpha - \beta p_2 - \gamma y)^2\right]\mathrm{d}t \tag{9}$$

Manufacturer's objective writes:

$$Maximize\, J^m = \int_0^\infty e^{-rt}\left[(p_1 - c)(\alpha - \beta p_2 - \gamma y) - c_1\left(\alpha - \beta p_2 - \gamma y - \overline{U}\right)^2 - ev^2 - fy^2\right]\mathrm{d}t \tag{10}$$

$$Subject\ to\ \dot{y} = \overline{U}(l - v) - \delta y \tag{11}$$

$$y(0) = y_0 > 0, \quad \text{where} \quad 0 \le v < l \quad \text{and} \quad p_2 \ge 0 \tag{12}$$

Model 3 is different from Model 2 in that the manufacturer and the retailer coordinate with each other closely as if they were a single company, i.e., it is the cooperative supply chain. Now there is only one objective function to be maximized by the integrated decision-maker combining the manufacturer and the retailer. The objective function for the cooperative supply chain writes:

$$J = \int_0^\infty e^{-rt}\left[(p_2 - c)(\alpha - \beta p_2 - \gamma y) - c_1\left(\alpha - \beta p_2 - \gamma y - \overline{U}\right)^2 - ev^2 - fy^2 - c_2(\alpha - \beta p_2 - \gamma y)^2\right]\mathrm{d}t$$

which is the combination of (9) and (10) after canceling out p_1.

Model 3

$$\begin{aligned} Maximize\, J = \int_0^\infty e^{-rt}\Big[&(p_2 - c)(\alpha - \beta p_2 - \gamma y) - c_1\left(\alpha - \beta p_2 - \gamma y - \overline{U}\right)^2 \\ &- ev^2 - fy^2 - c_2(\alpha - \beta p_2 - \gamma y)^2\Big]\mathrm{d}t \end{aligned} \tag{13}$$

$$Subject\,to\,\dot{y} = \overline{U}(l - v) - \delta y \tag{14}$$

$$y(0) = y_0 > 0, \quad \text{where} \quad 0 \le v < l \quad \text{and} \quad p_2 \ge 0 \tag{15}$$

Finally, Model 4 is different from Model 3 in that the consumers are ignorant of and insensitive to the manufacturer's pollution stock. Therefore, the demand function is now independent of the pollution stock, i.e., now the demand function is $(\alpha - \beta p_2)$ instead of $(\alpha - \beta p_2 - \gamma y)$.

Model 4

$$Maximize\,J = \int_0^\infty e^{-rt}\left[(p_2 - c)(\alpha - \beta p_2) - c_1\left(\alpha - \beta p_2 - \overline{U}\right)^2 - ev^2 - fy^2 - c_2(\alpha - \beta p_2)^2\right]dt \tag{16}$$

$$Subject\,to\,\dot{y} = \overline{U}(l - v) - \delta y \tag{17}$$

$$y(0) = y_0 > 0, \quad \text{where} \quad 0 \le v < l \quad \text{and} \quad p_2 \ge 0 \tag{18}$$

We present the detailed solution procedure for Model 2 in Appendix "Analysis for Model 2": Note that the solutions for other models are similar. We summarize the analysis results of the long-term equilibrium for the four models in Table 6.

2.2 Theorems

Based on the analysis of the differential game models, we develop theorems for the long-term equilibrium, i.e., the long-term equilibrium behaviors of the factors that determine the firm's dynamic decision-making to reduce the pollution.

Theorem 3 *At the long-term equilibrium, the manufacturer's effort to reduce its pollution* (v) *and the accumulated pollution* (y) *are identical for Model 1 and Model 4. That is,* $v^I_{LR} = v^{IV}_{LR}$ *and* $y^I_{LR} = y^{IV}_{LR}$.

Proof See Appendix "Proof of Theorem 3".

Theorem 3 implies that when the consumers are ignorant of and insensitive to the manufacturer's pollution emission, whether the supply chain is cooperative or competitive does not make any difference to the manufacturer's effort to reduce pollution and the ensuing accumulated pollution level.

Table 6 Summary of long-term equilibrium solutions

Model	Variables	Long-run equilibrium
Model 1	v_{LR}^{I}	$-\frac{\overline{U}}{2e}K_{11}$
	y_{LR}^{I}	$-\frac{1}{2f}K_{11}(r+\delta)$
	$p_{2\,LR}^{I}$	$\frac{\alpha(1+2\beta c_2)+\beta p_1}{2\beta(1+\beta c_2)}$
Model 2	v_{LR}^{II}	$-\frac{1}{\overline{U}}\left(\delta K_{21}-\overline{U}l\right)$
	y_{LR}^{II}	K_{21}
	$p_{2\,LR}^{II}$	$\frac{\alpha(1+2\beta c_2)+\beta p_1}{2\beta(1+\beta c_2)}-\frac{\gamma(1+2\beta c_2)}{2\beta(1+\beta c_2)}K_{21}$
Model 3	v_{LR}^{III}	$-\frac{1}{\overline{U}}\left(\delta K_{32}-\overline{U}l\right)$
	y_{LR}^{III}	K_{32}
	$p_{2\,LR}^{III}$	$\frac{\alpha(1+2\beta c_1+2\beta c_2)-2\beta c_1\overline{U}+\beta c}{2\beta(1+\beta c_1+\beta c_2)}-\frac{\gamma(1+2\beta c_1+2\beta c_2)}{2\beta(1+\beta c_1+\beta c_2)}K_{32}$
Model 4	v_{LR}^{IV}	$-\frac{\overline{U}}{2e}K_{41}$
	y_{LR}^{IV}	$-\frac{1}{2f}K_{41}(r+\delta)$
	$p_{2\,LR}^{IV}$	$\frac{\alpha+\beta c+2\beta c_1\left(\alpha-\overline{U}\right)+2\alpha\beta c_2}{2\beta(1+\beta c_1+\beta c_2)}$
Constants	K_{11}	$-\frac{2ef\overline{U}l}{f\overline{U}^2+e\delta(r+\delta)}$
	K_{21}	$\frac{\overline{U}^2\left\{\gamma(1+\beta c_2)\left[-(p_1-c)+2c_1\left(\alpha-\overline{U}\right)\right]-\gamma c_1(\alpha+2\alpha\beta c_2+\beta p_1)\right\}+2el\overline{U}(r+\delta)(1+\beta c_2)}{2(1+\beta c_2)\left[f\overline{U}^2+e\delta(r+\delta)\right]+\gamma^2\overline{U}^2c_1}$
	K_{31}	$\frac{4\beta(1+\beta c_1+\beta c_2)\left[f\overline{U}^2+e\delta(r+\delta)\right]-\gamma^2\overline{U}^2}{4e\beta(1+\beta c_1+\beta c_2)}$
	K_{32}	$\frac{4\beta el\overline{U}(1+\beta c_1+\beta c_2)(r+\delta)-\gamma\overline{U}^2\left(\alpha-\beta c+2\beta c_1\overline{U}\right)}{4\beta(1+\beta c_1+\beta c_2)\left[f\overline{U}^2+e\delta(r+\delta)\right]-\gamma^2\overline{U}^2}$
	K_{41}	$-\frac{2ef\overline{U}l}{f\overline{U}^2+e\delta(r+\delta)}$

Theorem 4 *There exists a transfer price*

$$\tilde{p}_1=\frac{\left[f\overline{U}^2+e\delta(r+\delta)\right]\left\{(1+\beta c_2)\left[c+2c_1\left(\alpha-\overline{U}\right)\right]-\alpha c_1(1+2\beta c_2)\right\}-el\gamma\overline{U}_{c_1}(r+\delta)}{(1+\beta c_1+\beta c_2)\left[f\overline{U}^2+e\delta(r+\delta)\right]}$$
$$\text{such that } y_{LR}^{II}<y_{LR}^{I}=y_{LR}^{IV}\text{ and } v_{LR}^{II}>v_{LR}^{I}=v_{LR}^{IV}\text{ if } p_1>\tilde{p}_1 \tag{19}$$

Proof See Appendix "Proof of Theorem 4".

Theorem 4 examines the pollution dynamics of the consumer-aware competitive supply chain. In a competitive supply chain, how to set a transfer price is an important mechanism to coordinate various decisions of the manufacturer and the retailer within the supply chain.

Rearranging Eq. (19) yields

$$\tilde{p}_1 = \left(\frac{1+\beta c_2}{1+\beta c_1+\beta c_2}\right)c - \left(\frac{2\overline{U}(1+\beta c_2)-\alpha}{1+\beta c_1+\beta c_2}\right)c_1 - \frac{el\gamma\overline{U}c_1(r+\delta)}{(1+\beta c_1+\beta c_2)\left[f\overline{U}^2+e\delta(r+\delta)\right]},$$

we infer that $\tilde{p}_1$ is less than the unit production cost c when $\overline{U}$ is large compared with other parameters. Consequently, $p_1 > \tilde{p}_1$ holds given sufficiently large $\overline{U}$, since it is reasonable to assume that the manufacturer would charge a transfer price p_1 to the retailer that is higher than its unit production cost c, i.e., $p_1 > c > \tilde{p}_1$. Therefore, we infer that when the capacity is relatively large, the long-term cumulative pollution of Model 2, the consumer-aware competitive supply chain is smaller than that of Model 1 and Model 4. Similarly, we deduce that when the capacity is relatively large, the manufacturer's long-term effort to reduce pollution in Model 2, the consumer-aware competitive supply chain is larger than that in Model 1 and Model 4. That is, under a normal situation, the manufacturer makes more effort to reduce pollution when the consumers are aware of and sensitive to the pollution stock and the supply chain is competitive than when the consumers are ignorant of the pollution.

Theorem 5 *There exists a market potential level*

$$\tilde{\alpha} = \frac{el\gamma\overline{U}(r+\delta)}{\left[f\overline{U}^2+e\delta(r+\delta)\right]} - \beta\left(2c_1\overline{U}-c\right) \text{ such that } y_{LR}^{III} < y_{LR}^{I} = y_{LR}^{IV} \text{ and } v_{LR}^{III} > v_{LR}^{I} = v_{LR}^{IV} \text{ if } \alpha > \tilde{\alpha}. \tag{20}$$

Proof See Appendix "Proof of Theorem 5".

Theorem 5 investigates the consumer-aware cooperative supply chain in terms of pollution emission, according to the potential market size. The potential market size directly influences the payoff in the supply chain, thus having a huge impact on the decisions including optimal emission and abatement in the supply chain.

In addition, from the long-term equilibrium derived in Table 6, we know that the long-term demand in Model 3 remains positive if and only if $\alpha > \frac{el\gamma\overline{U}(r+\delta)}{\left[f\overline{U}^2+e\delta(r+\delta)\right]} - \beta\left(2c_1\overline{U}-c\right) = \tilde{\alpha}$. Therefore, we infer that in general $\alpha > \tilde{\alpha}$ holds and the long-term cumulative pollution of Model 3, the consumer-aware cooperative supply chain is smaller than that of Model 1 and Model 4.

Similarly, we deduce that in general the manufacturer's long-term effort to reduce pollution in Model 3, the consumer-aware cooperative supply chain is larger than that in Model 1 and Model 4. That is, under a normal situation, the manufacturer makes more effort to reduce pollution when the consumers are aware of and sensitive to the pollution stock and the supply chain is cooperative than when the consumers are ignorant of the pollution.

Theorem 6 *It holds that* $\frac{\partial y^i_{LR}}{\partial f} < 0$ and $\frac{\partial v^i_{LR}}{\partial f} > 0$, where i = I, II, III, IV.

Proof See Appendix "Proof of Theorem 6".

Theorem 6 implies that as the government's penalty on the manufacturer's pollution stock increases, the manufacturer makes more effort to reduce the pollution and therefore the pollution stock decreases. Another route from the government's penalty to the reduction in the pollution stock is more direct, i.e., the government's penalty directly affects the manufacturer's profit function.

2.3 Numerical Examples

To visualize the analysis outcomes, we conduct numerical analysis. Parameter values for the base case are shown in Table 7. These values are based on the smartphone manufacturing industry: (1) the transfer price of a smartphone is about \$500 per unit ($p_1 = 50$), and the unit production cost of a smartphone is about \$200 per unit ($c = 20$); (2) one unit of manufacturer capacity emits 0.01 ton of pollution ($l = 0.01$); (3) the potential market size is about 2,000,000 units per month ($\alpha = 200,000$). Accordingly, the major variables are interpreted as follows: the measure of manufacturer's pollution accumulation (y) is 10 tons, the measure of sales price is \$10 per unit, and the measure of manufacturer's pollution abatement effort (v) is ton per unit capacity.

Table 8 summarizes the long-term equilibrium values of the variables and profits for the four models. It shows that Model 1 and Model 4 end up with the largest pollution accumulation (y) along with the least effort to reduce the pollution (v). This is in general consistent with the first three theorems in the previous section. That is, in terms of pollution reduction, the consumer-ignorant supply chain performs much worser than the consumer-aware supply chain does. But purely from the consumer welfare's perspective, the consumer-ignorant cooperative supply chain, i.e., Model 4, performs much better than others except for Model 3, the consumer-aware cooperative supply chain. That is, under Model 4 and Model 3, the equilibrium sales price is the lowest and the market demand per period is the largest. This result is consistent with the literature, i.e., the cooperative supply chain is better for the market since it eliminates double marginalization. Similarly, the two cooperative supply chains, i.e., Model 3 and Model 4, perform better in terms of the supply chain profits. That is, under Model 3 and Model 4, the supply chain profit,

Table 7 Parameter values for the numerical analysis

r	p_1	c	c_1	c_2	e	f	l	δ	α	β	γ	$\overline{U}$
0.004	50	20	10^{-5}	10^{-5}	10^9	0.01	0.01	0.1	200,000	2,000	0.1	100,000

Table 8 Long-term equilibrium for the base case

Variable	Model 1	Model 2	Model 3	Model 4
Pollution accumulation (y)	942.0	801.5	763.6	942.0
Pollution abatement effort (v)	0.00906	0.00920	0.00924	0.00906
Sales price (p_2)	75.49	75.47	60.56	60.58
Market demand per period	49,020	48,980	78,809	78,846
Supply chain profit for the entire period	6.44942×10^8	6.44102×10^8	7.59763×10^8	7.60593×10^8
Manufacturer profit for the entire period	3.38570×10^8	3.38215×10^8	–	–
Retailer profit for the entire period	3.06373×10^8	3.05887×10^8	–	–

which sums the retailer's profit and the manufacturer's, is the largest. Again this result is consistent with the literature, which supports that the cooperative supply chain generates more profit for the entire supply chain than the competitive supply chain does. Of course, this does not imply that Model 3 and Model 4 are fairer for the supply chain partners than Model 1 and Model 2. Whether the cooperative supply chain is fairer or not depends on how to share the increased profit between supply chain partners. Now which supply chain is the best? First, from the consumer welfare's perspective and also from the supply chain profit's perspective, Model 3 and Model 4 are better than the others. In addition, between Model 3 and Model 4, Model 3 is more desirable then Model 4, since it generates the least amount of pollution.

Let us consider more details of the numerical analysis. Figure 6 illustrates the dynamics of cumulative pollution and pollution abatement effort in the consumer-aware competitive supply chain compared with the consumer-ignorant supply chains. In Fig. 6, there are two regions: Region 1 ($p_1 > \tilde{p}_1$) implies that the consumer-aware competitive supply chain emits smaller pollution and invests more in pollution abatement than the consumer-ignorant supply chains, and Region 2 ($p_1 < \tilde{p}_1$) shows the reverse. But as shown in the figure, the consumer-aware competitive supply chain outperforms the consumer-ignorant supply chains in terms of environmental performance in most cases (i.e., Region 1 is much larger than Region 2), where the transfer price and the plant capacity are sufficient enough to guarantee a positive payoff for the manufacturer. It is unlikely to observe Region 2, where the consumer-aware competitive supply chain leads to a larger emission and a smaller abatement effort than the consumer-ignorant supply chains do, except for some extreme cases, where the plant capacity is very small compared to the market size and the transfer price is also very low.

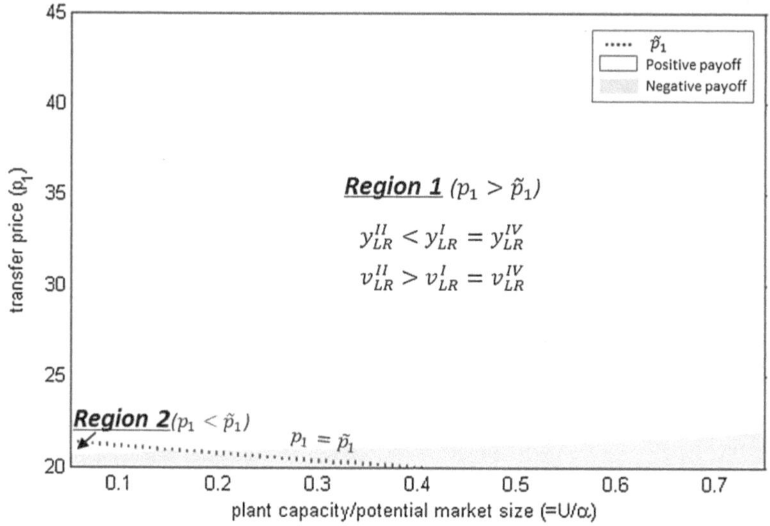

Fig. 6 Pollution dynamics of consumer-aware competitive versus consumer-ignorant supply chain. *Note* Region 1 implies $p_1 > \tilde{p}_1$, thus $y_{LR}^{II} < y_{LR}^{I} = y_{LR}^{IV}$ and $v_{LR}^{II} > v_{LR}^{I} = v_{LR}^{IV}$ hold. In contrast, Region 2 implies $p_1 < \tilde{p}_1$, thus $y_{LR}^{II} > y_{LR}^{I} = y_{LR}^{IV}$ and $v_{LR}^{II} < v_{LR}^{I} = v_{LR}^{IV}$ hold

Figure 7 compares the consumer-aware cooperative supply chain and the consumer-ignorant supply chains. It shows two regions, i.e., one with the potential market size larger than $\tilde{\alpha}$ (Region 1) and the other with the potential market size smaller than $\tilde{\alpha}$ (Region 2). Consistent with Theorem 5, the cumulative pollution is smaller and the pollution abatement effort is larger in the consumer-aware cooperative supply chain than the consumer-ignorant supply chains, provided that the potential market size is large enough to yield a positive payoff and also a positive demand for the supply chain. Intuitively, the cooperative supply chain earns a larger profit by eliminating double marginalization, thus allocating more resources to the abatement effort when the consumers are aware of the firm's environmental performance. When the potential market size is smaller than $\tilde{\alpha}$, the consumer-aware cooperative supply chain might emit more pollution and invest less in the abatement than the consumer-ignorant supply chains. But this circumstance is not sustainable, since such a low potential market size would ultimately put the supply chain out of business in the long term, it would not be profitable for the firm to be in such a market.

Comparing Model 2 and Model 3, Fig. 8 shows how the consumer's environmental awareness influences the firm's pollution abatement effort and cumulative pollution for the consumer-aware supply chains. As proved in Theorem 3, in the consumer-ignorant supply chains, the supply chain coordination (i.e., competitive vs. cooperative) has little to do with the firm's abatement effort and pollution emission. In the consumer-aware supply chains, however, it has a significant impact on the pollution-related decisions—the manufacturer in the competitive supply

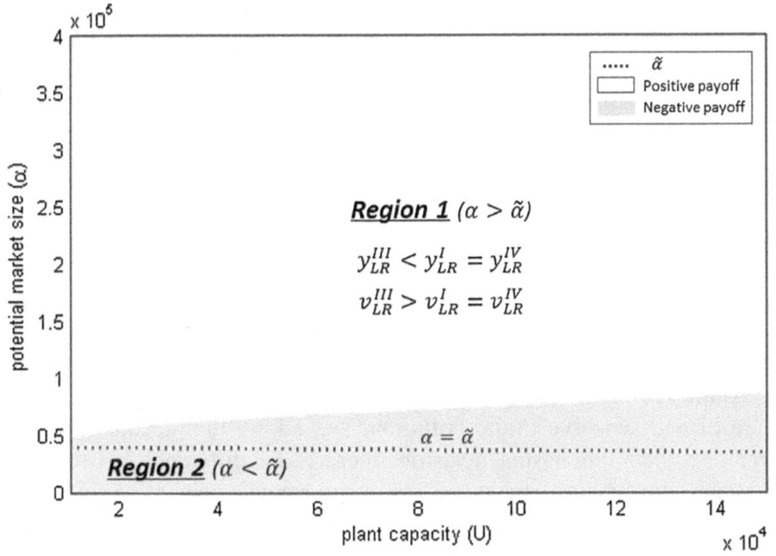

Fig. 7 Pollution dynamics of consumer-aware cooperative versus consumer-ignorant supply chain. *Note* Region 1 implies $\alpha > \tilde{\alpha}$, thus $y_{LR}^{III} < y_{LR}^{I} = y_{LR}^{IV}$ and $v_{LR}^{III} > v_{LR}^{I} = v_{LR}^{IV}$ hold. In contrast, Region 2 implies $\alpha < \tilde{\alpha}$, thus $y_{LR}^{III} > y_{LR}^{I} = y_{LR}^{IV}$ and $v_{LR}^{III} < v_{LR}^{I} = v_{LR}^{IV}$ hold

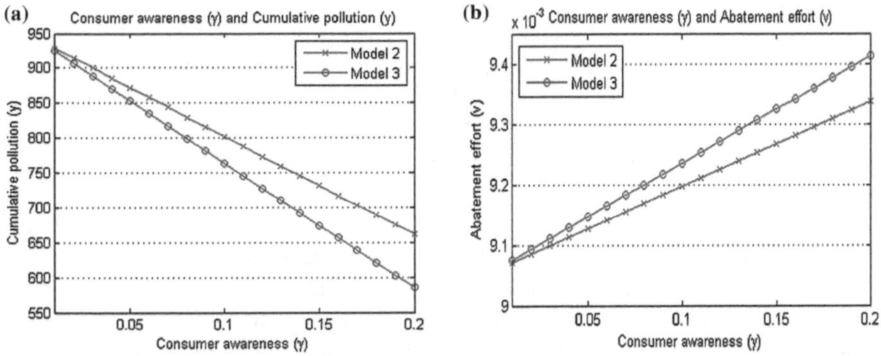

Fig. 8 Long-term pollution-related equilibrium depending on the consumer's awareness

chain invests less in its abatement activity and thus emits more pollution than in the cooperative supply chain. But Fig. 8 indicates that the effect of supply chain coordination is relatively insignificant when the consumer's environmental awareness is low. That is, even in the consumer-aware supply chains, the supply chain coordination results in a significant difference in terms of the firm's emission and abatement behaviors, only when the consumer awareness is sufficiently high. For instance, in Fig. 8a, the cumulative pollution (y) of Model 3 is about 98% of that of Model 2 when the consumer awareness is relatively low, e.g., $\gamma = 0.05$, whereas it is less than 93% when the consumer awareness is relatively high, e.g., $\gamma = 0.15$. This strongly confirms the important role played by the consumer's environment awareness in curbing the firm's pollution emission.

2.4 Discussion and Conclusion

In this section, we further elaborate on each theorem. Theorem 3 highlights the importance of consumer awareness in reducing the pollution emitted by the manufacturer. It is well known that whether the supply chain participants are competing or cooperating with each other affects the consequences of supply chain strategy to a great extent. However, our analysis strongly indicates that unless the consumers are aware of and sensitive to the pollution, i.e., taking into account the pollution when making their purchasing decision, there is no difference between the two supply chain coordination strategies, competitive and cooperative, in influencing the manufacturer to reduce its pollution emission.

Theorem 4 puts forth that a sufficiently large transfer price from the retailer to the manufacturer ensures that the consumer-aware competitive case is better than the consumer-ignorant cases in reducing the pollution. It also implies that if the transfer price is excessively low, it gives the manufacturer little incentive to make an effort to reduce the pollution, leading to less investment in pollution abatement effort and therefore more accumulated pollution stock.

Theorem 5 shows that the consumer-aware cooperative case is better than the consumer-ignorant cases as long as the potential market size is sufficiently large. By proving that such a condition should hold in order for the long-term demand to be positive, it actually confirms that the consumer-aware cooperative case is always better than the consumer-ignorant ones in reducing the pollution, under normal market conditions.

Finally, Theorem 6 clearly demonstrates that the government penalty is effective for each of the four cases, implying that when executed properly, the government penalty can play an important and effective role in curbing the environmental degradation due to the pollution emission in the supply chain.

We have conducted a numerical analysis to visualize the implications of the theorems. The numerical outcomes are consistent with the theorems. One intriguing observation is concerned with the comparison between the consumer-aware cooperative and the consumer-aware competitive supply chain. As expected, the cooperative supply chain performs better than the competitive counterpart. But the magnitude of its advantage enlarges as the level of consumer awareness increases. As discussed already, this observation further highlights the key role played by the consumer awareness in reducing the pollution emission.

Our analysis offers significant economic as well as managerial insights. First of all, Theorem 3 sheds light on understanding the essential role played by the consumers in controlling the pollution in a supply chain. It makes a potentially significant contribution to the literature by postulating a theory, which seems to contradict other existing ones in the literature. That is, the literature suggests that a cooperative supply chain performs better than a competitive supply chain or at least that the two supply chains, competitive and cooperative, generate different outcomes in most cases. But, our first theorem puts forth that the two supply chains, competitive and cooperative, are not different in terms of pollution reduction, unless the consumers are fully aware of the harmful effect of the pollution and take it into account when making their purchasing decisions. Stated differently, an elimination of double marginalization in a supply chain improves supply chain efficiency in terms of profit, but it alone cannot be a complete solution to environmental issues in a supply chain if it is not accompanied by proper change in the broader society (e.g., mindset of consumers). Therefore, reducing pollution and improving environmental sustainability in a supply chain should be approached from a perspective of entire economic system than the supply chain only.

It is also insightful that the transfer price determines whether the consumer-aware competitive case performs better in reducing the pollution than the consumer-ignorant cases. We conjecture in most situations the transfer price is high enough to make the consumer-aware competitive case perform better than the consumer-ignorant cases. Nevertheless, it is not impossible for the consumer-ignorant cases to perform better than the consumer-aware competitive case if the transfer price is excessively low, and thus the manufacturer has little motivation to make an effort to reduce the pollution. This also alludes the retailer's role to curb pollution of upstream manufacturer in the supply chain. For instance, a retailer might be able to influence manufacturer's payoffs and incentives for abatement by negotiating the transfer price with the manufacturer. This role of retailers, or intermediaries in general, would have more impact when upstream manufacturers do not fully take into account the end consumer's awareness in their decisions. For example, a supplier that provides LCD panel to a set manufacturer might not be well aware of the end consumer's awareness (e.g., the consumer cares about the carbon footprint of a supply chain). In such circumstances, the intermediary (the set manufacturer here) might impact the upstream supplier's incentive mechanism via negotiating the transfer price. We believe that the role of these intermediaries in improving environmental sustainability in supply chains deserves a closer examination in the future research.

For the consumer-aware cooperative case, the analysis result is much stronger. Although the analysis shows that the potential market size determines whether the consumer-aware cooperative case performs better than the consumer-ignorant cases, it turns out that in order to have a feasible solution, i.e., under any possible realistic situations, the consumer-aware cooperative case always performs better in curbing the pollution than the consumer-ignorant cases.

Finally, it is intriguing to note that the government penalty always forces the firm to increase its investment in pollution abatement and thus to reduce the accumulated pollution stock, regardless of whether the supply chain is competitive or cooperative and also whether the consumers are aware or ignorant. In this light, the government policy seems to be an effective tool to control the pollution. Nevertheless, this observation does not necessarily mean that the government penalty is a tool more effective than the consumer's environmental awareness. In fact, it can be a good topic for future research.

The analysis outcome helps both government policy makers and managers understand the complicated dynamics among critical factors related with consumers and supply chain strategies so as for them to make a decision to control the pollution more effectively. In essence, it strongly suggests that to reduce the pollution, managers and policy makers should try two methodologies simultaneously, one to encourage more coordination in the supply chain and the other to educate and motivate consumers to be more active for the environmental causes.

Exercise Problems

1. Solve Model 3 in Sect. 2 in this chapter to get the open-loop steady-state Nash equilibrium strategies as in Table 6.
2. A firm sells its product at the market price p and the market demand is fixed as Q. Producing one unit of product incurs a units of pollutants, and the government imposes a pollution penalty faQ in total. Let $x(t)$ be the firm's cumulative investment in emission-reducing technologies up to t, i.e., $x(t) = \int_0^t u(t)\mathrm{d}t$ and $x(0) = x_0$. Also, it is expected that a would decrease in x, i.e., $a = a_0 x^m$ $(m < 0)$. Assume that the cost for unit investment is c, and the maximum level of investment that the firm can afford is k at t. The firm wants to determine the optimal investment strategy $u(t)$ that maximizes its profit over the planning period $0 \le t \le T$. We assume a short enough T so that the discounting effect can be disregarded.

   (a) Formulate the optimal control problem.
   (b) Discuss the optimal investment strategy in emission-reducing technologies.

3. A pollution stock $y(t)$ increases proportionally with sales quantity of a manufacturing firm. The pollution stock can be reduced by abatement efforts $u(t)$ of the firm, and the dynamic equation for the pollution stock is $\dot{y}(t) = D(t) - u(t)$. Consumer demand depends on the sales price p and the pollution stock $y(t)$, according to $D(t) = \alpha - \beta p - \gamma y(t)$, where the coefficient γ represents the consumers' environmental awareness. Abatement efforts cost $\rho u^2/2$ to the firm. The sales price p is exogenously given. The firm decides $u(t)$ to maximize its profit over the fixed planning period $0 \le t \le T$. We assume that the initial pollution stock at $t = 0$ is zero and the discounting effect can be disregarded.

   (a) Derive the terminal pollution stock $y(T)$ of the firm.
   (b) Discuss the impact of increasing consumers' environmental awareness (i.e., increasing γ) on the terminal pollution stock of the firm.

4. Suppose a supply chain consists of one retailer and one manufacturer. Manufacturer's production activity entails a pollution stock $y(t)$ which increases proportionally with the production quantity. The pollution stock can be reduced by abatement efforts $v(t)$ of the manufacturer and the dynamic equation for the pollution stock is $\dot{y}(t) = D(t) - v(t)$. Consumer demand depends on the sales price $p(t)$ and the pollution stock $y(t)$, according to the function $D(t) = \alpha - \beta p(t) - \gamma y(t)$. Because the pollution stock also influences the retailer's profit via the consumer demand, the retailer supports a fixed portion $\phi \in [0, 1]$ of manufacturer's abatement costs, $\frac{ev^2}{2}$. The wholesale price w is exogenously given, and the unit production cost of the manufacturer is normalized to zero. The retailer decides $p(t)$ and the manufacturer decides $v(t)$ over the fixed planning period $0 \leq t \leq T$. We assume that the discounting effect can be disregarded. Show that the manufacturer's open-loop Nash equilibrium abatement efforts $v(t)$ increase with the retailer's cost-sharing portion ϕ.
5. Solve the following differential game model and suggest the stationary (long run) equilibrium of the state variable y, assuming open-loop Nash Equilibrium strategies.

$$Max_{p_2} J^r = \int_0^\infty e^{-rt}[(p_2 - p_1 - hv)(\alpha - \beta p_2)]dt$$

$$Max_v J^m = \int_0^\infty e^{-rt}\left[(p_1 + hv)(\alpha - \beta p_2) - ev^2 - fy^2\right]dt$$

Subject to $\dot{y} = \overline{U}(l - v) - \delta y,\ y(0) = y_0.$

6. Assume that the government penalty f is proportional to the increment of pollution, that is, the total pollution penalty imposed on a firm is $f\dot{y}$. A firm's pollution stock y increases according to $\dot{y} = q - v$ and $y(0) = y_0$, where q is a production quantity and v is an abatement effort. The cost associated with the firm's abatement effort is v^2, and the market price p decreases with the production quantity, i.e., $p = a - q$. The unit production cost is c. The firm wants to decide optimal p and v that maximize its profit over $t \in [0, T]$.
   (a) Formulate an optimal control problem of the firm, using r as a discounting rate.
   (b) Show that the optimal pricing and abatement strategies are time-independent.

Appendix 1

The Hamiltonian function is given by:

$$H = (p - c)(\alpha - \beta p - \gamma y) - c_1\left(\alpha - \beta p - \gamma y - \overline{U}\right)^2 - ev^2 \\ - fy^2 + \lambda\left[\overline{U}(l - v) - \delta y\right]$$

Assuming interior solutions, we obtain from the optimality conditions:

$$p = \frac{(1 + 2\beta c_1)(\alpha - \gamma y) - 2\beta c_1 \overline{U} + \beta c}{2\beta(1 + \beta c_1)}, \tag{21}$$

$$v = -\frac{\lambda \overline{U}}{2e}. \tag{22}$$

The solutions that satisfy the necessary conditions are optimal. The objective function is concave in (v, p). All constraints are linear in (v, p).

Costate equation, using (21), is as follows:

$$\dot{\lambda} = (r + \delta)\lambda + \left[2f + 2\gamma^2 c_1 - \frac{\gamma^2(1 + 2\beta c_1)^2}{2\beta(1 + \beta c_1)}\right] y \\ + \left\{\frac{\gamma(1 + 2\beta c_1)\left[\alpha(1 + 2\beta c_1) - 2\beta c_1 \overline{U} + \beta c\right]}{2\beta(1 + \beta c_1)} - \left[\gamma c + 2\gamma c_1\left(\alpha - \overline{U}\right)\right]\right\} \tag{23}$$

From (22) and the state equation,

$$\lambda = \frac{2e}{\overline{U}^2}\left(\dot{y} + \delta y - \overline{U}l\right). \tag{24}$$

Substituting (24) into (23) and solving the second-order differential equation of y yield:

$$y(t) = A_1 e^{m_1 t} + A_2 e^{m_2 t} + K_2, \tag{25}$$

where

$$m_1 = \frac{r + \sqrt{r^2 + 4\delta(r + \delta) + \frac{\overline{U}^2}{e}\left[4f - \frac{\gamma^2}{\beta(1 + \beta c_1)}\right]}}{2} > r,$$

$$m_2 = \frac{r - \sqrt{r^2 + 4\delta(r + \delta) + \frac{\overline{U}^2}{e}\left[4f - \frac{\gamma^2}{\beta(1 + \beta c_1)}\right]}}{2} < 0.$$

$$K_1 = \left\{ \delta(r+\delta) + \frac{\overline{U}^2}{2e}\left[2f - \frac{\gamma^2}{2\beta(1+\beta c_1)}\right]\right\} = \frac{4e\beta\delta(1+\beta c_1)(r+\delta) + 4f\overline{U}^2\beta(1+\beta c_1) - \gamma^2\overline{U}^2}{4e\beta(1+\beta c_1)}$$

$$K_2 = \frac{4\beta e l\overline{U}(1+\beta c_1)(r+\delta) - \gamma\overline{U}^2(\alpha - \beta c + 2\beta c_1\overline{U})}{4\beta(1+\beta c_1)\left[f\overline{U}^2 + e\delta(r+\delta)\right] - \gamma^2\overline{U}^2}$$

Note that $K_1 > 0$ would hold under the reasonable ranges of parameters, assuming a positive market demand.

From (24) and (25), $A_1 = 0$ is obtained to guarantee that the limiting transversality condition $\lim_{T\to\infty} e^{-rT}\lambda(T) = 0$ holds under all parameters. Also, $A_2 = y_0 - K_2$.

Considering $m_2 < 0$, the long-run equilibrium solutions in Table 2 are readily determined from (21), (22), and (25).

Proof of Theorem 1

$$\frac{\partial y_{LR}}{\partial f} = -\frac{4\beta(1+\beta c_1)\overline{U}^2 K_2}{\left\{4\beta(1+\beta c_1)\left[f\overline{U}^2 + e\delta(r+\delta)\right] - \gamma^2\overline{U}^2\right\}}.$$

Since K_1 is assumed to be positive and K_2 is nonnegative, $\mathrm{sgn}\left(\frac{\partial y_{LR}}{\partial f}\right) < 0$.

Similarly, $\frac{\partial v_{LR}}{\partial f} = \frac{4\beta\delta(1+\beta c_1)\overline{U}}{\left\{4\beta(1+\beta c_1)\left[f\overline{U}^2 + e\delta(r+\delta)\right] - \gamma^2\overline{U}^2\right\}} \cdot K_2 > 0.$

Also it holds that $\frac{\partial^2 y_{LR}}{\partial f^2} = \frac{32\beta^2(1+\beta c_1)^2\overline{U}^4}{\left\{4\beta(1+\beta c_1)\left[f\overline{U}^2 + e\delta(r+\delta)\right] - \gamma^2\overline{U}^2\right\}^2} \cdot K_2 > 0$ and $\frac{\partial^2 v_{LR}}{\partial f^2} = -\frac{\delta}{\overline{U}} \cdot \frac{\partial^2 y_{LR}}{\partial f^2} < 0$

Proof of Theorem 2

$$\begin{aligned}\frac{\partial^2 y_{LR}}{\partial f \partial \gamma} &= \frac{\partial}{\partial \gamma}\left(\frac{\partial y_{LR}}{\partial f}\right) \\ &= \frac{4\beta(1+\beta c_1)\overline{U}^2}{\left\{4\beta(1+\beta c_1)\left[f\overline{U}^2 + e\delta(r+\delta)\right] - \gamma^2\overline{U}^2\right\}^3} \\ &\quad \times \left[3\overline{U}^4(\alpha - \beta c + 2\beta c_1\overline{U})\gamma^2 - 16\beta e l\overline{U}^3(1+\beta c_1)(r+\delta)\gamma \right. \\ &\quad \left. + 4\beta\overline{U}^4(1+\beta c_1)(\alpha - \beta c + 2\beta c_1\overline{U})f + 4\beta e\delta\overline{U}^2(r+\delta)(1+\beta c_1)(\alpha - \beta c + 2\beta c_1\overline{U})\right]\end{aligned}$$

It is easily shown that $\frac{\partial^2 v_{LR}}{\partial f \partial \gamma} = -\frac{\delta}{\overline{U}} \cdot \frac{\partial^2 y_{LR}}{\partial f \partial \gamma}$.

Let $g(\gamma,f) = A\gamma^2 - B\gamma + Cf + D$, where $A = 3\overline{U}^4(\alpha - \beta c + 2\beta c_1 \overline{U}) > 0$, $B = 16\beta el\overline{U}^3(1+\beta c_1)(r+\delta) > 0$, $C = 4\beta\overline{U}^4(1+\beta c_1)(\alpha - \beta c + 2\beta c_1\overline{U}) > 0$, $D = 4\beta e\delta\overline{U}^2(r+\delta)(1+\beta c_1)(\alpha - \beta c + 2\beta c_1\overline{U}) > 0$.

Note that $\text{sgn}\left(\frac{\partial^2 y_{LR}}{\partial f \partial \gamma}\right) = -\text{sgn}\left(\frac{\partial^2 v_{LR}}{\partial f \partial \gamma}\right) = \text{sgn}[g(\gamma,f)]$ hold as $K_1 > 0$.

Therefore,

if $(\gamma,f) \in \{(\gamma,f)|g(\gamma,f) = A\gamma^2 - B\gamma + Cf + D < 0\}$, $\frac{\partial^2 y_{LR}}{\partial f \partial \gamma} < 0$, and $\frac{\partial^2 v_{LR}}{\partial f \partial \gamma} > 0$;
if $(\gamma,f) \in \{(\gamma,f)|g(\gamma,f) = A\gamma^2 - B\gamma + Cf + D \geq 0\}$, $\frac{\partial^2 y_{LR}}{\partial f \partial \gamma} \geq 0$, and $\frac{\partial^2 v_{LR}}{\partial f \partial \gamma} \leq 0$.

Let $h(\gamma) = -\frac{A}{C}\gamma^2 + \frac{B}{C}\gamma - \frac{D}{C} = -\frac{A}{C}\left(\gamma - \frac{B}{2A}\right)^2 + \frac{B^2}{4AC} - \frac{D}{C}$.

Let $G = B^2 - 4AD = 16\beta e\overline{U}^6(r+\delta)(1+\beta c_1)\ [16\beta el^2(1+\beta c_1)(r+\delta) - 3\delta(\alpha - \beta c + 2\beta c_1\overline{U})^2]$. G determines whether $h(\gamma) = 0$ has real root(s) and the sign of maximum value of $h(\gamma)$. Note that $g(\gamma,f) \gtreqless 0$ is equivalent to $f \gtreqless h(\gamma)$ and $h(\gamma)$ has a concave form.

To summarize, if $G \leq 0$, $g(\gamma,f) \geq 0$ for all feasible f; if $G > 0$, $g(\gamma,f) \geq 0$ for $f \geq h(\gamma)$; $g(\gamma,f) < 0$ for $f < h(\gamma)$. Note that

$$\begin{aligned}\text{sgn}(G) &= \text{sgn}\left[16\beta el^2(1+\beta c_1)(r+\delta) - 3\delta(\alpha - \beta c + 2\beta c_1\overline{U})^2\right] \\ &= -\text{sgn}\left[\alpha - \left(\sqrt{\frac{16\beta el^2(1+\beta c_1)(r+\delta)}{3\delta}} + \beta c - 2\beta c_1\overline{U}\right)\right].\end{aligned}$$

Also, assuming $G > 0$, the real roots of $h(\gamma) = 0$ are strictly positive as $A, B, D > 0$.

Appendix 2

Analysis for Model 2

We present the solution procedure for Model 2 in detail and omit others, since they are similar to that for Model 2. Recall **Model 2**:

$$\text{Maximize } J^r = \int_0^\infty e^{-rt}\left[(p_2 - p_1)(\alpha - \beta p_2 - \gamma y) - c_2(\alpha - \beta p_2 - \gamma y)^2\right]\text{d}t$$

$$\text{Maximize } J^m = \int_0^\infty e^{-rt}\left[(p_1 - c)(\alpha - \beta p_2 - \gamma y) - c_1(\alpha - \beta p_2 - \gamma y - \overline{U})^2 - ev^2 - fy^2\right]\text{d}t$$

$$\text{Subject to } \dot{y} = \bar{U}(l - v) - \delta y$$
$$y(0) = y_0 > 0, \text{ where } 0 \leq v < l \text{ and } p_2 \geq 0.$$

The Hamiltonian for the manufacturer's problem is

$$H^m = (p_1 - c)(\alpha - \beta p_2 - \gamma y) - c_1\left(\alpha - \beta p_2 - \gamma y - \overline{U}\right)^2 - ev^2 \\ - fy^2 + \lambda_1\left[\overline{U}(l - v) - \delta y\right] \tag{26}$$

Assuming interior solutions, necessary conditions for optimality lead to

$$v = -\frac{\lambda_1 \overline{U}}{2e} \tag{27}$$

$$p_1 = \begin{cases} 0 \\ p_1^s \in [0, M] \text{ if } \alpha - \beta p_2 - \gamma y \begin{Bmatrix} < \\ = \\ > \end{Bmatrix} 0 \\ M \end{cases} \tag{28}$$

Optimality condition for p_1 implies that p_1^* is a bang-bang policy, with the possibility of a singular control whenever $\alpha - \beta p_2 - \gamma y = 0$ holds. Note that $p_1 = 0$ can be excluded when $\alpha - \beta p_2 - \gamma y$, the market demand, is nonnegative. If the market demand is zero, we may assign p_1 an arbitrary value (e.g., M) because p_1 is indeterminate, and the choice of p_1 does not affect the Hamiltonian. Furthermore, the optimal p_1 is M for any positive market demand. Therefore, we can expect that the manufacturer always sets his maximum price at M (upper bound of p_1) as long as the market demand remains nonnegative. (see Jørgensen 1986)

The Hamiltonian for the retailer's problem is

$$H^r = (p_2 - p_1)(\alpha - \beta p_2 - \gamma y) - c_2(\alpha - \beta p_2 - \gamma y)^2 + \lambda_2\left[\overline{U}(l - v) - \delta y\right] \tag{29}$$

Assuming interior solutions, a necessary condition for optimality yields

$$p_2 = \frac{(1 + 2\beta c_2)(\alpha - \gamma y) + \beta p_1}{2\beta(1 + \beta c_2)}. \tag{30}$$

The solutions that satisfy the necessary conditions are optimal. The objective function of the manufacturer is concave in (v, p_1), and the objective function of retailer is concave in p_2. All constraints are linear in (v, p_1, p_2).

Costate equations are, using (30) as follows:

$$\dot{\lambda}_1 = (r + \delta)\lambda_1 + \left(\frac{\gamma^2 c_1}{1 + \beta c_2} + 2f\right)y + \gamma(p_1 - c) \\ - 2\gamma c_1\left(\alpha - \overline{U}\right) + \frac{\gamma c_1[\alpha(1 + 2\beta c_2) + \beta p_1]}{(1 + \beta c_2)}, \tag{31}$$

$$\dot{\lambda}_2 = (r+\delta)\lambda_2 + \left[2\gamma^2 c_2 - \frac{\gamma^2(1+2\beta c_2)^2}{2\beta(1+\beta c_2)}\right]y + \frac{\gamma(1+2\beta c_2)[\alpha(1+2\beta c_2)+\beta p_1]}{2\beta(1+\beta c_2)} \\ - \gamma(p_1+2\alpha c_2). \tag{32}$$

Using (11) and (27) yields $\lambda_1 = \frac{2e}{\overline{U}^2}\left(\dot{y}+\delta y-\overline{U}l\right)$. Therefore, after rearranging, (31) becomes

$$\ddot{y} - r\dot{y} - \left[\delta(r+\delta) + \frac{\overline{U}^2}{2e}\left(\frac{\gamma^2 c_1}{1+\beta c_2}+2f\right)\right]y \\ = \frac{\overline{U}^2}{2e}\left[\gamma(p_1-c) - 2\gamma c_1\left(\alpha-\overline{U}\right) + \frac{\gamma c_1(\alpha+2\alpha\beta c_2+\beta p_1)}{(1+\beta c_2)}\right] - (r+\delta)\overline{U}l. \tag{33}$$

Corresponding homogenous equation is $\ddot{y} - r\dot{y} - \left[\delta(r+\delta) + \frac{\overline{U}^2}{2e}\left(\frac{\gamma^2 c_1}{1+\beta c_2}+2f\right)\right] y = 0$, and the auxiliary equation is $m_2^2 - rm_2 - \left[\delta(r+\delta) + \frac{\overline{U}^2}{2e}\left(\frac{\gamma^2 c_1}{1+\beta c_2}+2f\right)\right] = 0$, where the two roots are

$$m_{21} = \frac{r+\sqrt{r^2+4\left[\delta(r+\delta)+\frac{\overline{U}^2}{2e}\left(\frac{\gamma^2 c_1}{1+\beta c_2}+2f\right)\right]}}{2} > r, \\ m_{22} = \frac{r-\sqrt{r^2+4\left[\delta(r+\delta)+\frac{\overline{U}^2}{2e}\left(\frac{\gamma^2 c_1}{1+\beta c_2}+2f\right)\right]}}{2} < 0. \tag{34}$$

A particular solution of y from (33) is

$$K_{21} = \frac{\overline{U}^2\{\gamma(1+\beta c_2)[-(p_1-c)+2c_1(\alpha-\overline{U})] - \gamma c_1(\alpha+2\alpha\beta c_2+\beta p_1)\} + 2el\overline{U}(r+\delta)(1+\beta c_2)}{2(1+\beta c_2)\left[f\overline{U}^2+e\delta(r+\delta)\right]+\gamma^2\overline{U}^2 c_1}. \tag{35}$$

Therefore, a general solution of y is $y(t) = A_{21}e^{m_{21}t} + A_{22}e^{m_{22}t} + K_{21}$ which yields

$$\lambda_1 = \frac{2e}{\overline{U}^2}\left(m_{21}A_{21}e^{m_{21}t} + m_{22}A_{22}e^{m_{22}t} + \delta A_{21}e^{m_{21}t} + \delta A_{22}e^{m_{22}t} + \delta K_{21} - \overline{U}l\right). \tag{36}$$

Also, to guarantee that the limiting transversality condition $\lim_{T\to\infty} e^{-rT}\lambda_1(T) = 0$ holds for all parameters, $A_{21} = 0$ ($\because m_{21} - r > 0$).

Therefore,

$$y(t) = A_{22}e^{m_{22}t} + K_{21} \tag{37}$$

$$\lambda_1(t) = \frac{2e}{\overline{U}^2}\left[(m_{22} + \delta)A_{22}e^{m_{22}t} + \delta K_{21} - \overline{U}l\right] \tag{38}$$

$$v(t) = -\frac{\lambda_1 \overline{U}}{2e} = -\frac{1}{\overline{U}}\left[(m_{22} + \delta)A_{22}e^{m_{22}t} + \delta K_{21} - \overline{U}l\right] \tag{39}$$

$$p_2(t) = \frac{(1+2\beta c_2)(\alpha - \gamma y) + \beta p_1}{2\beta(1+\beta c_2)} = \frac{\alpha(1+2\beta c_2) + \beta p_1}{2\beta(1+\beta c_2)} - \frac{\gamma(1+2\beta c_2)}{2\beta(1+\beta c_2)}y \tag{40}$$

Calculating the coefficient using the initial condition gives $A_{22} = y_0 - K_{21}$, and the long-term equilibriums are as follows:

$$y_{LR}^{II} \to K_{21} \tag{41}$$

$$v_{LR}^{II} \to -\frac{1}{\overline{U}}\left(\delta K_{21} - \overline{U}l\right) \tag{42}$$

$$p_{2\,LR}^{II} \to \frac{\alpha(1+2\beta c_2) + \beta p_1}{2\beta(1+\beta c_2)} - \frac{\gamma(1+2\beta c_2)}{2\beta(1+\beta c_2)}K_{21}. \tag{43}$$

Proof of Theorem 3

From Table 6, it is obvious that $K_{11} = K_{41}$ holds, which leads to $v_{LR}^{I} = v_{LR}^{IV}$ and $y_{LR}^{I} = y_{LR}^{IV}$.

Proof of Theorem 4

Since $y_{LR}^{I} = y_{LR}^{IV}$ and $v_{LR}^{I} = v_{LR}^{IV}$ is proved in Theorem 3, we examine $y_{LR}^{IV} - y_{LR}^{II}$ and $v_{LR}^{II} - v_{LR}^{IV}$ only.

First,

$$y_{LR}^{IV} - y_{LR}^{II} = -\frac{1}{2f}K_{41}(r+\delta) - K_{21} \tag{44}$$

Plugging the values of K_{41} and K_{21} in Table 6 into (44) and rearranging the equation, $y_{LR}^{IV} - y_{LR}^{II} > 0$ is equivalent to

$$(1+\beta c_1+\beta c_2)\left[f\overline{U}^2+e\delta(r+\delta)\right]p_1 > \left[f\overline{U}^2+e\delta(r+\delta)\right]\{(1+\beta c_2)[c+2c_1(\alpha-\overline{U})] \\ -\alpha c_1(1+2\beta c_2)\} - el(r+\delta)\gamma\overline{U}c_1. \tag{45}$$

Since $(1+\beta c_1+\beta c_2)\left[f\overline{U}^2+e\delta(r+\delta)\right]$ is positive, (45) is equivalent to

$$p_1 > \frac{\left[f\overline{U}^2+e\delta(r+\delta)\right]\{(1+\beta c_2)[c+2c_1(\alpha-\overline{U})]-\alpha c_1(1+2\beta c_2)\}-el\gamma\overline{U}c_1(r+\delta)}{(1+\beta c_1+\beta c_2)\left[f\overline{U}^2+e\delta(r+\delta)\right]} = \tilde{p}_1. \tag{46}$$

Therefore, it holds that $y_{LR}^{II} < y_{LR}^{I} = y_{LR}^{IV}$, when $p_1 > \tilde{p}_1$.

Similarly,

$$v_{LR}^{II} - v_{LR}^{IV} = -\frac{1}{\overline{U}}\left(\delta K_{21} - \overline{U}l\right) + \frac{\overline{U}}{2e}K_{41}. \tag{47}$$

Plugging the values of K_{21} and K_{41} in Table 6 into (47) and rearranging the equation, $v_{LR}^{II} - v_{LR}^{IV} > 0$ is equivalent to $p_1 > \tilde{p}_1$.

Therefore, it holds that $v_{LR}^{II} > v_{LR}^{I} = v_{LR}^{IV}$, when $p_1 > \tilde{p}_1$.

Proof of Theorem 5

Let $Q_1 = \left[f\overline{U}^2+e\delta(r+\delta)\right]$ and $Q_2 = 4\beta(1+\beta c_1+\beta c_2)\left[f\overline{U}^2+e\delta(r+\delta)\right] - \gamma^2\overline{U}^2$.

Note that Q_2 is positive since K_{31} in Table 6 is assumed to be positive.

Since $y_{LR}^{I} = y_{LR}^{IV}$ and $v_{LR}^{I} = v_{LR}^{IV}$ is proved in Theorem 3, we examine $y_{LR}^{IV} - y_{LR}^{III}$ and $v_{LR}^{III} - v_{LR}^{IV}$ only.

First,

$$y_{LR}^{IV} - y_{LR}^{III} = -\frac{1}{2f}K_{41}(r+\delta) - K_{32} \tag{48}$$

Plugging the values of K_{41} and K_{32} in Table 6 into (48) and utilizing Q_1 and Q_2, $y_{LR}^{IV} - y_{LR}^{III} > 0$ is equivalent to

$$\gamma\overline{U}Q_1\alpha > el(r+\delta)\gamma^2\overline{U}^2 - \beta\gamma\overline{U}\left[f\overline{U}^2 + e\delta(r+\delta)\right]\left(2c_1\overline{U} - c\right) \tag{49}$$

Since $\gamma\overline{U}Q_1$ is positive, (49) is equivalent to

$$\alpha > \frac{el(r+\delta)\gamma\overline{U}}{\left[f\overline{U}^2 + e\delta(r+\delta)\right]} - \beta\left(2c_1\overline{U} - c\right) = \tilde{\alpha}. \tag{50}$$

Therefore, it holds that $y_{LR}^{III} < y_{LR}^{I} = y_{LR}^{IV}$, when $\alpha > \tilde{\alpha}$.
Similarly,

$$v_{LR}^{III} - v_{LR}^{IV} = -\frac{1}{\overline{U}}\left(\delta K_{32} - \overline{U}l\right) + \frac{\overline{U}}{2e}K_{41} \tag{51}$$

Plugging the values of K_{32} and K_{41} in Table 6 into (51) and rearranging the equation, $v_{LR}^{III} - v_{LR}^{IV} > 0$ is equivalent to $\alpha > \tilde{\alpha}$.
Therefore, it holds that $v_{LR}^{III} > v_{LR}^{I} = v_{LR}^{IV}$, when $\alpha > \tilde{\alpha}$.
Furthermore, since $D = \alpha - \beta p_2 - \gamma y$ in Model 3, the long-term demand is

$$\begin{aligned} D_{LR}^{III} &= \alpha - \beta p_{2\,LR}^{III} - \gamma y_{LR}^{III} \\ &= \alpha - \beta\left[\frac{\alpha(1+2\beta c_1+2\beta c_2) - 2\beta c_1\overline{U} + \beta c}{2\beta(1+\beta c_1+\beta c_2)} - \frac{\gamma(1+2\beta c_1+2\beta c_2)}{2\beta(1+\beta c_1+\beta c_2)}K_{32}\right] - \gamma K_{32} \\ &= \frac{1}{2(1+\beta c_1+\beta c_2)}\alpha + \frac{2\beta c_1\overline{U} - \beta c}{2(1+\beta c_1+\beta c_2)} - \frac{\gamma}{2(1+\beta c_1+\beta c_2)} \\ &\quad \times \frac{4\beta el\overline{U}(1+\beta c_1+\beta c_2)(r+\delta) - \gamma\overline{U}^2\left(\alpha - \beta c + 2\beta c_1\overline{U}\right)}{4\beta(1+\beta c_1+\beta c_2)\left[f\overline{U}^2 + e\delta(r+\delta)\right] - \gamma^2\overline{U}^2}. \end{aligned} \tag{52}$$

For the long-term demand to be positive, rearranging Eq. (52), $\alpha + 2\beta c_1\overline{U} - \beta c - \frac{4\beta\gamma el\overline{U}(1+\beta c_1+\beta c_2)(r+\delta) - \gamma^2\overline{U}^2\left(\alpha - \beta c + 2\beta c_1\overline{U}\right)}{4\beta(1+\beta c_1+\beta c_2)\left[f\overline{U}^2 + e\delta(r+\delta)\right] - \gamma^2\overline{U}^2} > 0$ should hold which is equivalent to

$$\left[f\overline{U}^2 + e\delta(r+\delta)\right]\alpha + \left(2\beta c_1\overline{U} - \beta c\right)\left[f\overline{U}^2 + e\delta(r+\delta)\right] - \gamma el\overline{U}(r+\delta) > 0. \tag{53}$$

Therefore, after rearranging (53), it is obvious that $D_{LR}^{III} > 0$ holds if and only if $\alpha > \frac{el(r+\delta)\gamma\overline{U}}{\left[f\overline{U}^2 + e\delta(r+\delta)\right]} - \beta\left(2c_1\overline{U} - c\right) = \tilde{\alpha}$.

Proof of Theorem 6

Part 1

$$\frac{\partial y_{LR}^{I}}{\partial f}=-\frac{\overline{U}^{2}}{\left[f\overline{U}^{2}+e\delta(r+\delta)\right]^{2}}\cdot e\overline{U}l(r+\delta)<0 \tag{54}$$

It is obvious that (54) holds because all parameters are positive.

Similarly,

$$\frac{\partial v_{LR}^{I}}{\partial f}=-\frac{fl\overline{U}^{4}}{\left[f\overline{U}^{2}+e\delta(r+\delta)\right]^{2}}+\frac{l\overline{U}^{2}}{f\overline{U}^{2}+e\delta(r+\delta)}=\frac{el\overline{U}^{2}\delta(r+\delta)}{\left[f\overline{U}^{2}+e\delta(r+\delta)\right]^{2}}>0. \tag{55}$$

Since $y_{LR}^{I}=y_{LR}^{IV}$ and $v_{LR}^{I}=v_{LR}^{IV}$, it also holds that $\frac{\partial y_{LR}^{IV}}{\partial f}<0$, $\frac{\partial v_{LR}^{IV}}{\partial f}>0$.

Part 2

$$\frac{\partial y_{LR}^{II}}{\partial f}=-\frac{2\overline{U}^{2}(1+\beta c_2)}{\left\{2(1+\beta c_2)\left[f\overline{U}^{2}+e\delta(r+\delta)\right]+\gamma^{2}\overline{U}^{2}c_1\right\}^{2}}\left\{\overline{U}^{2}\{\gamma(1+\beta c_2)[-(p_1-c)+2c_1(\alpha-\overline{U})]\right.$$
$$\left.-\gamma c_1(\alpha+2\alpha\beta c_2+\beta p_1)\}+2el\overline{U}(r+\delta)(1+\beta c_2)\right\} \tag{56}$$

Utilizing the value of K_{21} in Table 6, $\frac{\partial y_{LR}^{II}}{\partial f}<0$ is equivalent to:

$$-\frac{2\overline{U}^{2}(1+\beta c_2)}{\left\{2(1+\beta c_2)\left[f\overline{U}^{2}+e\delta(r+\delta)\right]+\gamma^{2}\overline{U}^{2}c_1\right\}}\cdot K_{21}<0 \tag{57}$$

Note that K_{21} is assumed to be nonnegative for feasible controls. Therefore, (57) holds.

Similarly,

$$\frac{\partial v_{LR}^{II}}{\partial f}=\frac{2\delta\overline{U}(1+\beta c_2)}{\left\{2(1+\beta c_2)\left[f\overline{U}^{2}+e\delta(r+\delta)\right]+\gamma^{2}\overline{U}^{2}c_1\right\}}\cdot K_{21}>0 \tag{58}$$

Part 3

$$\frac{\partial y_{LR}^{III}}{\partial f} = -\frac{4\beta(1+\beta c_1+\beta c_2)\overline{U}^2}{\left\{4\beta(1+\beta c_1+\beta c_2)\left[f\overline{U}^2+e\delta(r+\delta)\right]-\gamma^2\overline{U}^2\right\}^2} \times \left[4\beta el\overline{U}(1+\beta c_1+\beta c_2)(r+\delta)-\gamma\overline{U}^2\left(\alpha-\beta c+2\beta c_1\overline{U}\right)\right] \quad (59)$$

Utilizing the value of K_{32} in Table 6, $\frac{\partial y_{LR}^{III}}{\partial f} < 0$ is equivalent to

$$-\frac{4\beta(1+\beta c_1+\beta c_2)\overline{U}^2}{\left\{4\beta(1+\beta c_1+\beta c_2)\left[f\overline{U}^2+e\delta(r+\delta)\right]-\gamma^2\overline{U}^2\right\}} \cdot K_{32} < 0 \quad (60)$$

Note that $\left\{4\beta(1+\beta c_1+\beta c_2)\left[f\overline{U}^2+e\delta(r+\delta)\right]-\gamma^2\overline{U}^2\right\} > 0$ since K_{31} is assumed to be positive. Also, K_{32} is assumed to be nonnegative for feasible controls. Therefore, (60) holds.

Similarly,

$$\frac{\partial v_{LR}^{III}}{\partial f} = \frac{4\beta\delta(1+\beta c_1+\beta c_2)\overline{U}}{\left\{4\beta(1+\beta c_1+\beta c_2)\left[f\overline{U}^2+e\delta(r+\delta)\right]-\gamma^2\overline{U}^2\right\}} \cdot K_{32} > 0. \quad (61)$$

Appendix 3: Literature Review

In environmental economics, a number of studies examined how the regulators could effectively induce pollution reduction through diverse instruments and incentives (Benchekroun and van Long 1998; Chen and Sheu 2009; Jung et al. 1996; Krass et al. 2013; Li et al. 2014; Li 2013). Milliman and Prince (1989) investigated five regulatory regimes such as direct controls, emission subsidies, emission taxes, free marketable permits, and auctions marketable permits, and examined which policy would facilitate firms' technological change in the pollution control most effectively. Jung et al. (1996) also evaluated the effectiveness of various regulatory instruments in terms of firms' incentives to develop and adopt pollution abatement technology. Subramanian et al. (2007) studied how firms' pollution reduction strategies would vary under a regulator's decision on the permits for emissions.

More recently, in the literature, there have been emerging interests in operational and market factors beyond the regulatory policies in inducing firms' environmental performance. In reviewing studies in environmentally and socially sustainable operations, Tang and Zhou (2012) emphasized that the role of environmentally conscious consumers and cooperation within a supply chain deserves further investigation. Despite the importance of supply chain coordination and consumer's environmental awareness, however, how the two factors simultaneously affect firm's environmental performance remains largely unexplored. One notable exception is a study of Zhang et al. (2015). They examined how consumer's environmental awareness would influence the order quantity decision and profits in three supply chain scenarios, i.e., a centralized supply chain, a decentralized supply chain, and a decentralized supply chain with a return contract.

Regarding supply chain coordination, the commonly accepted view is that a cooperative supply chain leads to higher environmental performance and sustainability (Handfield et al. 1997; Hollos et al. 2012; Simpson 2010). Ni et al. (2010) found that socially responsible or environmental performance is the highest in the cooperative supply chain, where the supplier and the manufacturer jointly maximize the supply chain profit, mainly because the double marginalization problem is eliminated. Lou et al. (2015) also examined three supply chain configurations and found that the cooperative supply chain in which the manufacturer and the retailer act as a single firm and the supply chain coordinated by a revenue sharing contract invest in emission reduction more than the decentralized supply chain. Klassen and Vachon (2003) empirically showed that an increased collaboration in the supply chain helps the firms invest more in environmental programs.

Consumer's increasing preference for environment-friendly products is another important mechanism to motivate firms to reconsider their environmental strategy. Lee (2010) described how consumer's environmental awareness influenced Esquel, one of the leading suppliers of premium cotton, to improve its environmental sustainability. Several studies incorporated environmentally conscious consumers explicitly and analyzed its impact on firms' decisions and environmental performance (Conrad 2005; Du et al. 2015; Ghosh and Shah 2012; Wang et al. 2014). Bagnoli and Watts (2003) investigated how firms' competition for socially responsible or environment-friendly consumers influenced firm's decisions. Yalabik and Fairchild (2011) showed that pressures from environment-conscious consumers and regulators both lead to lower emissions as long as the initial emissions are not severe. Also, they found that high environmental competition between firms not only induces lower emissions but also improves the effectiveness of environmental pressures from consumers or regulators. Liu et al. (2012) also examined the impact of consumer's environmental awareness and firms' competition in production or retail on the supply chain.

Another important issue in modeling the firm's environmental effort is concerned with what actually generates pollution. For instance, is the pollution emission rate proportional to the production rate or production capacity? Examining firm's effort to reduce pollution, Subramanian et al. (2007) put forth two types of pollution reduction, one independent of and the other dependent on the production volume. Similarly, Chung et al. (2013) specified two sources of manufacturer's pollution emission, one due to the plant operations, e.g., the size of the capacity, independent of production rate and the other due to and proportional to the production rate. There is no shortage of empirical studies, which reported that the plant capacity is related with the firm's pollution emission rate, e.g., a plant with a larger capacity emits more pollution and thus has lower environmental performance (Grant et al. 2002; Gray and Shadbegian 2004; Laplante and Rilstone 1996; Ludwig 2004; Vachon and Klassen 2006). We conjecture that as long as the firm utilizes its capacity sufficiently, the firm's pollution emission rate is proportional to its plant capacity, which in turn is closely related to its production rate.

Erratum to: Optimal Control Applications for Operations Strategy

Bowon Kim

Erratum to:
B. Kim, *Optimal Control Applications for Operations Strategy*, DOI 10.1007/978-981-10-3599-9

In the original version of the book, the newly received corrections have to be incorporated in Chapters 1–5. The erratum book has been updated with the changes.

The updated online version of this book can be found at
http://dx.doi.org/10.1007/978-981-10-3599-9
http://dx.doi.org/10.1007/978-981-10-3599-9_1
http://dx.doi.org/10.1007/978-981-10-3599-9_2
http://dx.doi.org/10.1007/978-981-10-3599-9_3
http://dx.doi.org/10.1007/978-981-10-3599-9_4
http://dx.doi.org/10.1007/978-981-10-3599-9_5

B. Kim, *Optimal Control Applications for Operations Strategy*,
DOI 10.1007/978-981-10-3599-9_6

References

Agrawal VV, Ferguson M, Toktay LB, Thomas VM (2012) Is leasing greener than selling? Manag Sci 58:523–533
Alchian A (1963) Reliability of progress curves in airframe production. Econometrica 31:679–693
Arrow KJ (1962) The economic implications of learning by doing. Rev Econ Stud 29:155–173
Arrow KJ (1969) Classificatory notes on the production and transmission of technological knowledge. Am Econ Rev LIX(2):29–35
Bagnoli M, Watts SG (2003) Selling to socially responsible consumers: Competition and the private provision of public goods. J Econ Manag Strategy 12(3):419–445
Barros LL (1989) The optimization of repair decision using life-cycle cost parameters. IMA J Manag Math 9(4):403–413
Benchekroun H, van Long N (1998) Efficiency inducing taxation for polluting oliogopolists. J Public Econ 70:325–342
Bernstein F, Kök AG (2009) Dynamic cost reduction through process improvement in assembly networks. Manag Sci 55(4):552–567
Bertinelli L, Camacho C, Zou B (2014) Carbon capture and storage and transboundary pollution: a differential game approach. Eur J Oper Res 237:721–728
Bohn RE (1988) Learning by experimentation in manufacturing. Harvard Business School Working Paper #88-001
Bridges E, Yim CK, Briesch RA (1995) A high-tech product market share model with customer expectations. Mark Sci 14(1):61–81
Business Week (2004) Hyundai: Kissing Clunkers Goodbye—a five-year focus on quality has sent customer satisfaction soaring. Business Week (May 17, 2004)
Canbolat PG, Golany B, Rothblum UG (2012) A stochastic competitive research and development race where "winner takes all" with lower and upper bounds. J Optim Theory Appl 154:986–1014
Cellini R, Lambertini L (2009) Dynamic R&D with spillovers: competition vs cooperation. J Econ Dyn Control 33(3):568–582
Chanel O, Cleary S, Luchini S (2006) Does public opinion influence willingness-to-Pay? Evidence from the field. Appl Econ Lett 13(13):821–824
Chen X, Li L, Zhou M (2012) Manufacturer's pricing strategy for supply chain with warranty period-dependent demand. Omega 40:807–816
Chen YJ, Sheu J-B (2009) Environmental-regulation pricing strategies for green supply chain management. Transp Res Part E: Logist Transp Rev 45(5):667–677
Chintagunta PK, Rao VR (1996) Pricing strategies in a dynamic duopoly: a differential game model. Manag Sci 42(11):1501–1514
Chung SH, Weaver RD, Friesz TL (2013) Strategic response to pollution taxes in supply chain networks: dynamic, spatial, and organizational dimensions. Eur J Oper Res 231:314–327
Conrad K (2005) Price competition and product differentiation when consumers care for the environment. Environ Resour Econ 31(1):1–19

B. Kim, *Optimal Control Applications for Operations Strategy*,
DOI 10.1007/978-981-10-3599-9

Dai Y, Zhou SX, Xu Y (2012) Competitive and collaborative quality and warranty management in supply chains. Prod Oper Manag 21(1):129–144

Dockner E, Jørgensen S (1988) Optimal pricing strategies for new products in dynamic oligopolies. Mark Sci 7(4):315–334

Dockner E, Jorgensen S, Long NV, Sorger G (2000) Differential games in economics and management science. Cambridge University Press, United Kingdom

Dompere KK (1993) Technological progress and optimal supply price. Int J Prod Econ 32:365–381

Dorroh JR, Gulledge TR, Womer NK (1994) Investment in knowledge: a generalization of learning by experience. Manag Sci 40(8):947–958

Dosi G (1982) Technological paradigms and technological trajectories. Res Policy 11:147–162

Du S, Zhu J, Jiao H, Ye W (2015) Game-theoretical analysis for supply chain with consumer preference to low carbon. Int J Prod Res 53(12):3753–3768

Durham Y (2000) An experimental examination of double marginalization and vertical relationships. J Econ Behav Organ 42(2):207–229

El Ouardighi F, Benchekroun H, Grass D (2014) Controlling pollution and environmental absorption capacity. Ann Oper Res 220:111–133

El Ouardighi F, Jørgensen S, Pasin F (2013) A dynamic game with monopolist manufacturer and price-competing duopolist retailers. OR Spectr 35(4):1059–1084

Eliashberg J, Steinberg R (1987) Marketing-production decisions in an industrial channel of distribution. Manag Sci 33(8):981–1000

Eliashberg J, Steinberg R (1991) Competitive strategies for two firms with asymmetric production cost structures. Manag Sci 37(11):1452–1473

Feichtinger G, Dockner E (1985) Optimal pricing in a duopoly: a noncooperative differential games solution. J Optim Theory Appl 45(2):199–218

Feller I (1972) Production isoquants and the analysis of technological and technical change. Q J Econ LXXXVI(1):154–161

Fershtman C, Nitzan S (1991) Dynamic voluntary provision of public goods. Eur Econ Rev 35:1057–1067

Fine CH (1986) Quality improvement and learning in productive systems. Manag Sci 32 (10):1301–1315

Gaimon C, Ozkan GF, Napoleon K (2011) Dynamic resource capabilities: managing workforce knowledge with a technology upgrade. Organ Sci 22(6):1560–1578

Ghosh D, Shah J (2012) A comparative analysis of greening policies across supply chain structures. Int J Prod Econ 135(2):568–583

Goel RK (2006) Uncertain innovation with uncertain product durability. Appl Econ Lett 13 (13):829–834

Grant DS, Jones AW, Bergesen AJ (2002) Organizational size and pollution: the case of the U.S. chemical industry. Am Sociol Rev 67:389–407

Gray WB, Shadbegian RJ (2004) 'Optimal' pollution abatement—whose benefits matter, and how much? J Environ Econ Manag 47(3):510–534

Hamel G, Doz YL, Prahalad CK (1989) Collaborate with your competitors—and win. Harv Bus Rev 67(1):133–139

Handfield RB, Walton SV, Seegers LK, Melnyk SA (1997) Green' value chain practices in the furniture industry. J Oper Manag 15(4):293–315

Hartley JL, Zirger BJ, Kamath RR (1997) Managing the buyer-supplier interface for on-time performance in product development. J Oper Manag 15:57–70

Hicks JR (1932) The theory of wages. Macmillan, London

Hollos D, Blome C, Foerstl K (2012) Does sustainable supplier co-operation affect performance? Examining implications for the triple bottom line. Int J Prod Res 50(11):2968–2986

Iyer AV, Bergen ME (1997) Quick response in manufacturer retailer channels. Manag Sci 43 (4):559–570

Jaikumar R (1988) From filing and fitting to flexible manufacturing: A study in the evolution of process control. Harvard Business School Working Paper #88-045.

Jaikumar R, Bohn RE (1992) A dynamic approach to operations management: an alternative to static optimization. Int J Prod Econ 27:265–282

Jeuland AP, Shugan SM (1983) Managing channel profits. Mark Sci 2(2):239–272

Jørgensen S (1986) Optimal production, purchasing and pricing: a differential game approach. Eur J Oper Res 24(1):64–76

Jung C, Krutilla K, Boyd R (1996) Incentives for advanced pollution abatement technology at the industry level: an evaluation of policy alternatives. J Environ Econ Manag 30:95–111

Kamien MI, Schwartz NL (1978) Optimal exhaustible resource depletion with endogenous technical change. Rev Econ Stud 45:179–196

Kamien MI, Schwartz NL (1991) Dynamic optimization: the calculus of variations and optimal control in economics and management, 2nd edn. North-Holland

Klassen RD, Vachon S (2003) Collaboration and evaluation in the supply chain: the impact on plant-level environmental investment. Prod Oper Manag 12(3):336–352

Krass D, Nedorezov T, Ovchinikov A (2013) Environmental taxes and the choice of green technology. Prod Oper Manag 22:1035–1055

Laplante B, Rilstone P (1996) Empirical inspections and emissions of the pulp and paper industry in Quebec. J Environ Econ Manag 31:19–36

Lau AK, Tang E, Yam R (2010) Effects of supplier and customer integration on product innovation and performance: empirical evidence in Hong Kong manufacturers. J Prod Innovat Manag 27(5):761–777

Lee HL (2010) Don't tweak your supply chain: rethink it end to end. Harvard Bus Rev 88 (10):62–69

Leonard-Barton D (1992) The factory as a learning laboratory. Sloan Manag Rev 34(1)

Li J, Du W, Yang F, Hua G (2014) The carbon subsidy analysis in remanufacturing closed-loop supply chain. Sustainability 6(6):3861–3877

Li S (2013) Emission permit banking, pollution abatement and production-inventory control of the firm. Int J Prod Econ 146:679–685

Liu Z, Anderson TD, Cruz JM (2012) Consumer environmental awareness and competition in two-stage supply chains. Eur J Oper Res 218:602–613

Lou GX, Xia HY, Zhang JQ, Fan TJ (2015) Investment strategy of emission-reduction technology in a supply chain. Sustainability 7(8):10684–10708

Ludwig L (2004) The US acid rain program and its effect on SO2 emission levels. Issues in Polit Econ 13:11–22

Mahajan V, Muller E (1996) Timing, diffusion, and substitution of successive generations of technological innovations: the IBM mainframe case. Technol Forecast Soc Chang 51 (2):109–132

March JG, Olsen JP (1976) Ambiguity and choice in organizations. Universitets forlaget, Bergen, Norway

Milliman SR, Prince R (1989) Firm incentives to promote technological change in pollution control. J Environ Econ Manag 17:247–265

Miyazaki H (2009) An analysis of the relation between R&D and M&A in high-tech industries. Appl Econ Lett 16(2):199–201

Mody A (1989) Firm strategies for costly engineering learning. Manag Sci 35(4):496–512

Mukhopadhyay SK, Kouvelis P (1997) A differential game theoretic model for duopolistic competition on design quality. Oper Res 45(6):886–893

Nelson RR, Winter SG (1973) Toward an evolutionary theory of economic capabilities. Am Econ Rev 63(2):440–449

Ni D, Li KW, Tang X (2010) Social responsibility allocation in two-echelon supply chains: insights from wholesale price contracts. Eur J Oper Res 207:1269–1279

Park B-J, Srivastava MK, Gnyawali DR (2014) Walking the tight rope of coopetition: impact of competition and cooperation intensities and balance on firm innovation performance. Ind Mark Manag 43:210–221

Peng DX, Heim GR, Mallick DN (2014) Collaborative product development: the effect of project complexity on the use of information technology tools and new product development practices. Prod Oper Manag 23(8):1421–1438

Pennings E (2004) Optimal pricing and quality choice when investment in quality is irreversible. J Ind Econ 52(4):569–589

Pontyagin LS et al (1962) The mathematical theory of optimal processes. Translated from Russian by K.N. Trirogoff, Interscience, New York

Reinganum JF (1984) Practical implications of game theoretic models of R&D. Am Econ Rev 74:61–66

Rossana RJ (1985) Delivery lags and buffer stocks in the theory of investment by the firm. J Econ Dyn Control 9(2):153–193

Sáenz-Royo C, Salas-Fumás V (2013) Learning to learn and productivity growth: evidence from a new car-assembly plant. Omega 41:336–344

Schumpeter JA (1947) The creative response in economic history. J Econ Hist 7(2):149–159

Sengupta JK (2001) A model of Schumpeterian innovations. Appl Econ Lett 8(6):397–401

Shell K (1966) Toward a theory of inventive activity and capital accumulation. Am Econ Rev 5:62–68

Shen ZJM (2006) A profit-maximizing supply chain network design model with demand choice flexibility. Oper Res Lett 34(6):673–682

Simpson D (2010) Use of supply relationships to recycle secondary materials. Int J Prod Res 48 (1):227–249

Srivastava SK (2007) Green supply-chain management: a state-of-the-art literature review. Int J Manag Rev 9(1):53–80

Sterman JD, Repenning NP, Kofman F (1997) Unanticipated side effects of successful quality programs: exploring a paradox of organizational improvement. Manag Sci 43(4):503–521

Subramanian R, Gupta S, Talbot B (2007) Compliance strategies under permits for emissions. Prod Oper Manag 16(6):763–779

Tang CS, Zhou S (2012) Research advances in environmentally and socially sustainable operations. Eur J Oper Res 223(3):585–594

Tapiero CS (1987) Production learning and quality control. IEEE Trans 19(4):362–370

Thompson LG (1968) Optimal maintenance policy and sale date of a machine. Manag Sci 14 (9):543–550

Tidball M, Zaccour G (2009) A differential environmental game with coupling constraints. Optim Control Appl Methods 30(2):197–207

Tomiyama K (1984) Two-stage optimal control problems and optimality conditions. J Econ Dyn Control 9(3):317–337

Vachon S, Klassen RD (2006) Green project partnership in the supply chain: the case of the package printing industry. J Clean Prod 14(6):661–671

Von Hippel E (1994) "Sticky information" and the locus of problem solving: implications for innovation. Manag Sci 40(4):429–439

Wang K, Zhao Y, Cheng Y, Choi T-M (2014) Cooperation or competition? Channel choice for a remanufacturing fashion supply chain with government subsidy. Sustainability 6 (10):7292–7310

Wei J, Zhao J, Li Y (2015) Price and warranty period decisions for complementary products with horizontal firms' cooperation/noncooperation strategies. J Clean Prod 105:86–102

Weil D (1997) Implementing employment regulations: Insights on the determinants of regulatory performance. Industrial Relations Research Association

Wiener JL (1985) Are warranties accurate signals of product reliability? J Consum Res 12 (2):245–250

Wind J, Mahajan V (1997) Editorial: issues and opportunities in new product development: an introduction to the special issue. J Mark Res 34(1):1–12
Xiao W, Gaimon C (2013) The effect of learning and integration investment on manufacturing outsourcing decisions: theoretic approach. Prod Oper Manag 22(6):1576–1592
Yalabik B, Fairchild RJ (2011) Customer, regulatory, and competitive pressure as drivers of environmental innovation. Int J Prod Econ 131:519–527
Yelle LE (1979) The learning curve: historical review and comprehensive study. Decis Sci 10:302–328
Youssef SB, Breton M, Zaccour G (2013) Cooperating and non-cooperating firms in incentive and absorptive research. J Optim Theory Appl 157:229–251
Zhang L, Wang J, You J (2015) Consumer environmental awareness and channel coordination with two substitutable products. Eur J Oper Res 241(1):63–73
Zhao J (2000) An optimal quality cost model. Appl Econ Lett 7(3):185–188

Index

B. Kim, *Optimal Control Applications for Operations Strategy*,
DOI 10.1007/978-981-10-3599-9

FSC
www.fsc.org
MIX
Papier | Fördert
gute Waldnutzung
FSC® C083411

Kolibri 360 GmbH
Ferdinand-Jühlke-Straße 7
99095 Erfurt, Deutschland
produktsicherheit@kolibri360.de